Advances in Anatomy, Embryology and Cell Biology

Volume 239

AF802130

Advances in Anatomy, Embryology and Cell Biology publishes critical reviews and state-of-the-art research in the areas of anatomy, developmental and cellular biology. Founded in 1891, this book series has a long standing tradition of publishing focused and condensed information on a given topic with a special emphasis on biomedical and translational aspects.

The series is open to both contributed volumes (each collecting 7 to 15 focused reviews written by leading experts) and single-authored or multi-authored monographs (providing a comprehensive overview of their topic of research).

Advances in Anatomy, Embryology and Cell Biology is indexed in Medline, SCImago, SCOPUS.

Maria L. Golson
Editor

Molecular and Cell Biology of Pancreas Development, Function and Regeneration

Springer

Editor
Maria L. Golson
Division of Endocrinology, Diabetes and Metabolism
Johns Hopkins School of Medicine, Department of Medicine
Baltimore, MD, USA

ISSN 0301-5556 ISSN 2192-7065 (electronic)
Advances in Anatomy, Embryology and Cell Biology
ISBN 978-3-031-62231-1 ISBN 978-3-031-62232-8 (eBook)
https://doi.org/10.1007/978-3-031-62232-8

This Springer imprint is published by the registered company Springer Nature Switzerland AG
The registered company address is: Gewerbestrasse 11, 6330 Cham, Switzerland

Preface

Diabetes poses an escalating global health threat, affecting over 500 million individuals worldwide. The common factor underlying all types of diabetes is insufficient insulin production, which hinders peripheral tissues from absorbing or sequestering glucose. Insulin, vital for glucose regulation, is produced by β cells housed within pancreatic islets. The normal development of the pancreas, its islets, and β cells is essential for maintaining blood glucose homeostasis and averting diabetes and its associated complications. In addition, abnormalities in the exocrine pancreas also lead to diseases. For example, anomalies in ductal differentiation result in truncated ducts, leading to conditions like chronic pancreatitis, recurrent acute pancreatitis, and pancreatic insufficiency.

Advancements in sequencing and imaging technologies, new computational methods facilitating the integration and analysis of vast datasets, and new preclinical models have accelerated research into pancreas formation and function. This book concentrates on recent breakthroughs in understanding pancreas development, differentiation, and cell-fate maintenance. Its intended audience includes researchers who possess a fundamental grasp of this field and aspire to keep informed on its developments. The chapter authors are active researchers in their topics.

The first and second chapters delve into intrinsic mechanisms governing pancreas development. The first chapter elucidates the influence of epigenetics on pancreas development and its impact on β-cell function. These areas have benefited from enhanced comprehension of the histone code and sequencing technology. The second chapter explores the formation of pancreatic structure in three-dimensional space, focusing on ductal development. Advancements in imaging techniques have largely facilitated progress in this area.

The next three chapters scrutinize the role of the mature β cell in health and disease, along with attempts to restore functional β-cell mass in the face of β-cell failure. The third chapter examines attempts to replenish β-cell mass in diabetes via endogenous β-cell replication or transdifferentiating other pancreatic cell types. This effort has been enabled by an understanding of transcription factor and signaling networks gleaned from mouse models and ex vivo studies of human pancreatic tissue. The fourth delves into the knowledge of β-cell heterogeneity in health and disease, pushed by improvements in sorting live islet cells and single-cell technology. The fifth chapter discusses recent research regarding the amplifying pathways of insulin secretion, partly facilitated by breakthroughs in reporter molecules and imaging technology.

The sixth chapter delves into progress in δ-cell research, historically underexplored due to its rarity in human and mouse islets and a shared marker expression with neuronal and enteroendocrine cells. Advances in single-cell technologies, such as sequencing and imaging, and the development and implementation of unique mouse models and genetic reporters have accelerated research into this cell.

The last chapter describes the adverse effects of disrupted maternal metabolic health on β-cell development and function. Manipulations of preclinical models have driven research on this topic.

The work presented in these chapters demonstrates that new technologies have expanded the horizons of diabetes and pancreas research. Looking ahead, further technological advancement promises even greater progress in these fields. Continued investment in research and development, coupled with interdisciplinary collaboration, may bring breakthroughs in preventing, diagnosing, and treating diabetes and pancreatic disorders.

In closing, I want to thank the authors of this book and the publications referenced for their contributions to this body of work. Your dedication and efforts are vital for furthering pancreas and diabetes research.

Baltimore, MD — Maria L. Golson

Contents

Epigenetic Regulation of Pancreas Development and Function

Tanya Hans Pierre, Eliana Toren, Jessica Kepple, and Chad S. Hunter

Abstract The field of epigenetics broadly seeks to define heritable phenotypic modifications that occur within cells without changes to the underlying DNA sequence. These modifications allow for precise control and specificity of function between cell types—ultimately creating complex organ systems that all contain the same DNA but only have access to the genes and sequences necessary for their cell-type-specific functions. The pancreas is an organ that contains varied cellular compartments with functions ranging from highly regulated glucose-stimulated insulin secretion in the β-cell to the pancreatic ductal cells that form a tight epithelial lining for the delivery of digestive enzymes. With diabetes cases on the rise worldwide, understanding the epigenetic mechanisms driving β-cell identity, function, and even disease is particularly valuable. In this chapter, we will discuss the known epigenetic modifications in pancreatic islet cells, how they are deposited, and the environmental and metabolic contributions to epigenetic mechanisms. We will also explore how a deeper understanding of epigenetic effectors can be used as a tool for diabetes therapeutic strategies.

Keywords Islets · Epigenetics · Pancreas · Chromatin · Histone · Beta-cell · Diabetes

---

T. H. Pierre · E. Toren · J. Kepple · C. S. Hunter (✉)
Comprehensive Diabetes Center and Department of Medicine, Division of Endocrinology, Diabetes, and Metabolism, University of Alabama at Birmingham, Birmingham, AL, USA
e-mail: tpierre@uab.edu; etoren@uab.edu; jessica.kepple@ndm.ox.ac.uk; chadhunter@uabmc.edu; huntercs@uab.edu

M. L. Golson (ed.), *Molecular and Cell Biology of Pancreas Development, Function and Regeneration*, Advances in Anatomy, Embryology and Cell Biology 239, https://doi.org/10.1007/978-3-031-62232-8_1

Abbreviations

CTCF	CCCTC-binding factor
CBP	CREB-binding protein
DDR	DNA-damage response
DNMT	DNA methyltransferases
E	Embryonic day
HDT	Histone demethylases
HMT	Histone methyltransferases
JmjC	Jumonji domain-containing demethylases
MafA	MAF bZIP transcription factor A
MSI2	Musashi RNA binding protein 2
Paupar	Pax6 upstream antisense RNA
PRMT	Protein arginine methyltransferases
TET	ten-eleven translocation
TF	Transcription factor
T1D	Type 1 diabetes
ac	Acetylation
R	Arginine
BET	Bromodomains and Extra terminal domain
Cpa1	Carboxypeptidase-1
ceRNA	Competing endogenous RNA
DUBs	Deubiquitylases
DIO	Diet induced obesity
DSB	Double strand break
GWAS	Genome-wide association studies
GLP-1	Glucagon-like peptide 1
GSIS	Glucose-stimulated insulin secretion
HFD	High fat diet
H	Histone
HAT	Histone acetyltransferase
HDACs	Histone deacetylases
lncRNA	Long ncRNA
K	Lysine
MOF	Males absent on the first
MEG3	Maternally expressed gene 3
me	Methylation
miRISCs	miRNA-induced silencing complex
mRE	microRNA recognition element
Ngn3	Neurogenin-3
NAD	Nicotinamide adenine dinucleotide
ncRNA	Non-coding RNA
NOD	Nonobese diabetic
OGTT	Oral glucose tolerance test

Pdx1	Pancreatic duodenal homeobox 1
piRNA	piwi-interacting RNA
PRC	Polycomb repressive complex
ROIT	Regulator of insulin transcription
Rnf20	Ring finger 20
Rnf40	Ring finger 40
siRNA	Small interfering RNA
TSA	Trichostatin A
TrxG	Trithorax group
T2D	Type 2 diabetes
ub	Ubiquitination
Ucn3	Urocortin-3
5hmC	5-hydroxymethylcytosine
5mC	5-methylcytosine
βFaar	β-cell function and apoptosis regulator
βLINC1	β-cell Long Intergenic Noncoding RNA 19

1 Introduction

Genomic DNA sequences provide the genetic code that governs cellular identity and dictates the biological processes required for life. While DNA nucleotide base-pair sequences contain genomic instructions, the fine-tuning of cellular diversity and function is achieved through epigenetics. Epigenetics defines heritable modifications of gene expression that do not directly alter underlying DNA sequences. A primary mechanism of epigenetic control is the regulation of chromatin accessibility to transcription factor (TF) and coregulator complexes, which can be influenced by histone and DNA modifications. Non-coding RNAs can also carry out epigenetic control through modification of chromatin structure by regulating the time and place in which genomic regions are accessible. Additionally, overall chromatin topography, not just accessibility, is also manipulated in epigenetics. The placement of enhancers, promoters, and other genetic regulatory elements dictates their ability to interact with genes and enable transcription. This intricate and dynamic regulatory system allows organs, such as the pancreas, to develop and function with numerous diverse compartments and cell types, despite having the same genetic code in every nucleus.

The varied cell types of the pancreas, each with unique functions, make it an ideal environment for delving into epigenetic control, especially during the highly regulated orchestration of embryonic development. Pancreatic development begins around embryonic day (E) 9.5 in mice, with dorsal and ventral pancreatic buds emerging on either side of the developing endodermal gut tube (Jorgensen et al. 2007). The first TF to demarcate multipotent pancreatic progenitors is pancreatic duodenal homeobox 1 (PDX1), and subsequent progenitors are classified into pools of either "tip" or "trunk" cells, as defined by Carboxypeptidase1 (CPA1) and Neurogenin-3 (Ngn3) expression, respectively (Zhou et al. 2007; Jonsson et al.

1994). While trunk cells will subsequently populate the exocrine compartment of the pancreas, the NGN3$^+$ endocrine progenitors will differentiate into the five hormone-secreting cell types that populate the islet, including the insulin-producing β-cells, glucagon-producing α-cells, and somatostatin-producing δ-cells. Over time, PDX1 expression becomes enriched in β-cells, which require expression of the islet maturity TF marker MAFA (MAF bZIP transcription factor A), among others, for proper glucose sensing and insulin secretion (Zhang et al. 2005).

The primary responsibility of mature β-cells is to produce, package, and secrete insulin in response to glucose metabolism. In turn, insulin reduces blood glucose levels, protecting an organism from hyperglycemia. In addition to PDX1 and MAFA, the TFs NKX6.1, FOXA2, and NEUROD1 facilitate these functions of mature β-cells (Wortham and Sander 2021). Reduced expression of β-cell TFs or dysregulation of their transcriptional complexes, and thereby loss of β-cell function, disrupts glucose homeostasis and often results in the development of diabetes (Davidson et al. 2021). Type 2 diabetes (T2D) is characterized by the depletion of functional β-cell mass and peripheral insulin resistance and makes up the majority of global diabetes cases (Tinajero and Malik 2021). Briefly, during T2D, β-cell function intensifies to keep up with increasing glycemic demands from overnutrition or obesity. This increased demand can cause ER stress and unfolded protein response (UPR) in the β-cell. Chronic high blood glucose levels can have glucotoxic effects on the β-cell, which cause hypertrophy or hyperplasia that additionally contributes to dysregulation. Comprising only about 5% of diabetes cases, type 1 diabetes (T1D) results from autoimmune-mediated destruction of β-cells (Xie et al. 2020). Autoreactive T cells infiltrate the pancreas, activate inflammatory pathways, and subsequently trigger the destruction of insulin-producing β-cells. Both T1D and T2D are heritable diseases, but environmental factors that impact epigenetics and gene expression, including diet, the intrauterine environment, exercise, and viral exposure, can contribute to diabetes development and general β-cell dysfunction (Rewers and Ludvigsson 2016).

Ultimately, the combinatorial effect of TF actions with epigenetic effectors shape the diversity and specificity of pancreatic cell identity and function. In this chapter, we will define epigenetic modifications and their modifiers in the pancreas, identify environmental and metabolic contributions to epigenetic mechanisms, and explore how a deeper understanding of epigenetic effectors can be used as tools for various diabetes therapeutic strategies.

2 Types of Epigenetic Modifications and their Activities

2.1 DNA Modifications

2.1.1 Methylation and DNMTs

DNA methylation is a DNA modification that typically elicits gene silencing through the placement of a methyl group on a cytosine residue within a CpG dinucleotide. This addition changes the activity of the modified DNA segment without altering the base-pair sequence itself. DNA methyltransferases (DNMTs) are responsible for depositing methyl groups, while TET enzymes facilitate their removal. TETs can reverse DNMT action through the conversion of the 5-methylcytosine (5mC) methylation mark into a 5-hydroxymethylcytosine (5hmC) intermediate that precedes thymine DNA glycosylase-mediated base-excision repair (Luo et al. 2018) (Fig. 1).

Of the four primary classes of DNMTs, DNMT1 is a key regulator of pancreas development and maintenance of α-cell and β-cell identity. Tandem loss of DNMT1 and an α-cell TF, ARX, in mature α-cells results in a transition to a β-cell-like state, highlighting potential plasticity of α-cells as well as DNMT1 importance (Chakravarthy et al. 2017). While DNMT1 loss triggers β-cell to α-cell reprogramming upon β-cell-specific loss, ablation of DNMT1 alone in α- or δ-cells has no effect on cell fate (Damond et al. 2017). Embryonic ablation of DNMT1 solely in α-cells, from conception to weaning, triggers a progressive reduction in adult α-cells and by 12-months DNMT1-deficient α-cell mice have fivefold less α-cells than controls. Developmental loss of DNMT1, however, did not impact α-cell maturity or differentiation. Similarly, deletion of DNMT1 in δ-cells had no effect on their differentiation or survival.

While DNA methylation patterns are established primarily during early development and tissue differentiation, they are frequently disrupted in disease states. Differential methylation patterns have been identified prior to and during the onset of both T1D and T2D, which adds therapeutic interest to investigating and understanding this epigenetic mark. Methylation has been suggested as a potential biomarker of T1D, where methylation near the *INSULIN* locus was observed in T1D

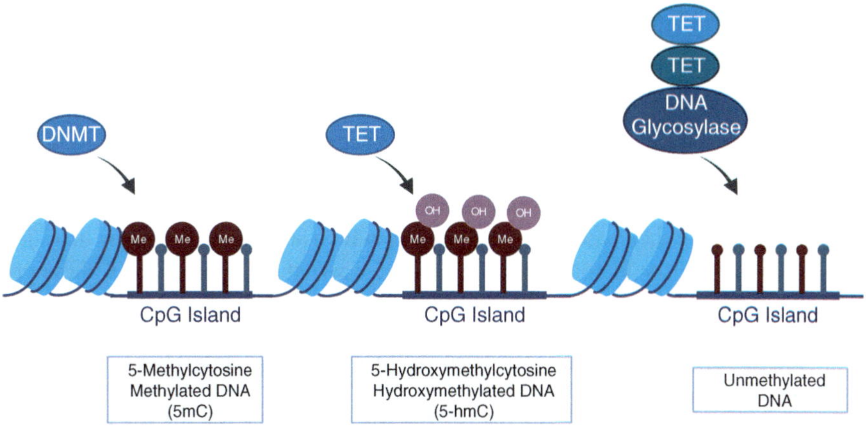

Fig. 1 DNMT and TET function. DNMT (DNA methyltransferase) enzymes place a methyl group (Me) on a cytosine (C) residue of a CpG dinucleotide in unmethylated DNA to create 5-methylcytosine (5mC) DNA. TET enzymes can reverse DNMT action by catalyzing the hydroxylation of 5mC into a 5-hydroxymethlcytosine (5hmC) via the addition of OH residues. This 5-hmC intermediate precedes base-excision repair by DNA glycosylases, which convert modified residues back to C, resulting in a return to the unmethylated state. Created with BioRender.com

patients (Carry et al. 2020). Interestingly, differentially methylated DNA, including *INSULIN*, may also be a biomarker of β-cell death (Liu et al. 2018). Longitudinal studies in T1D individuals showed differential methylation patterns in 10 genomic regions (primarily located near protein-coding genes) prior to islet-autoimmunity onset, suggesting that this pattern could be a precursor to T1D development (Johnson et al. 2020). As for T2D, hyperglycemia and high lipid exposure are associated with altered DNA methylation in pancreatic islets. Human islets incubated with high glucose demonstrated an overall increase in DNA methylation, including increased methylation of the *PDX1* distal promoter and enhancer regions (Hall et al. 2018). In a longitudinal study of individuals with T2D, peripheral blood samples were collected following an oral glucose tolerance test (OGTT), and T2D imparted increased differential methylation relative to control individuals throughout the study. Of the differentially methylated targets identified, *MSI2* (Musashi RNA binding protein 2) had a strong positive correlation with insulin sensitivity. MSI2 negatively regulates murine *insulin* expression, and its expression is responsive to T2D-like conditions of lipotoxicity and ER stress (Jeon et al. 2017).

2.2 Histone Modifications

Histones are the proteins around which eukaryotic DNA is tightly wrapped to allow extensive genomic material to fit within the nucleus. The DNA-histone complex forms chromatin, whose basic unit is the nucleosome (Fig. 2). Nucleosomes comprise an octamer of two of each of the histone proteins (histone(H) H2A, H2B, H3, H4) wrapped in 146 base pairs of DNA. The level of DNA compaction around histones can influence gene transcription, with more tightly wound DNA reducing accessibility of transcriptional machinery to genomic sequences. This compaction is controlled by post-translational amino- or carboxy-terminal histone tail modifications that influence downstream chromatin function (Fischle et al. 2003). The regulators that modify histone tails through the placement of modifications are called epigenetic writers. There are various types of epigenetic writers, including acetyltransferases, methyltransferases, ubiquitin ligases, and kinases. These epigenetic modifications are then recognized by readers, which are effector proteins that identify and interpret the histone marks left by each writer. Manipulation of these epigenetic modifications allows for multiple levels of cellular diversity, as marks encoded by writers can be dynamically reversed by their respective epigenetic erasers (Zhang et al. 2015). A list of these modifications and their general effects on transcription can be found in Table 1.

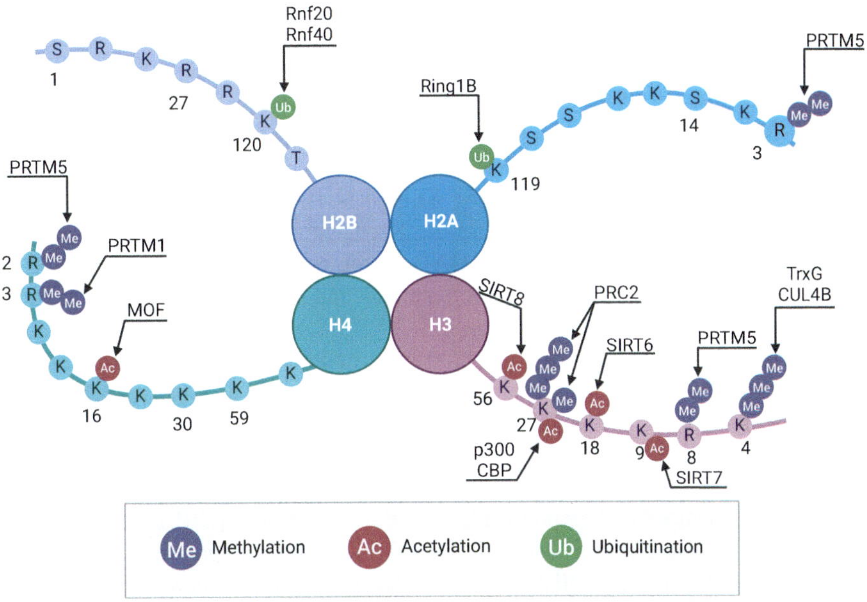

Fig. 2 Histones and their modifications. Histone octamer comprised two of each histone unit: H2A, H2B, H3, and H4, each with histone tails. These flexible histone tails can be post-translationally modified by various factors (black) that deposit methylation (purple), acetylation (red), or ubiquitination marks (green). These post-translational modifications (PTMs) affect DNA interactions by altering the charge on histone proteins and thereby tightly regulating nucleosome stability, gene accessibility, and gene expression. Created with BioRender.com

Table 1 Histone modifications and their general effects

Residue	Modifier	Transcriptional effect
Histone acetylation		
H3K27Ac	CBP and p300	Activation
H4K16Ac	MOF	Activation
H3K18Ac	SIRT6	Activation
H3K9Ac	SIRT7	Activation
H3K56Ac	SIRT8	Activation
Histone methylation		
H3K27me	PRC2	Activation
H3K27me3	PRC2	Repression
H3K4me3	TrxG and CUL4B	Activation
H4R3me2a	PRTM1	Activation
H2AR3me2s	PRTM5	Repression
H3R8me2s	PRTM5	Repression
H4R2me2s	PRTM5	Repression
Histone Ubiquitination		
H2AK119ub H2AK119ub	Ring1B Bmi1	Repression
H2BK120ub1	Rnf20 and Rnf40	Activation and repression

2.2.1 Histone Acetylation

HATs

Histone acetylation is a modification associated with open chromatin and, therefore, gene activation. Histone acetyltransferases (HATs) carry out acetylation by placing acetyl groups on the lysine (K) residues of histones. HATs can be classified into two main groups based on where they are localized within cells, with class A HATs localized to the nucleus and class B HATs to the cytoplasm (Gujral et al. 2020). Below, we detail various HATs in the pancreas and studies that have led to their discovery and characterization thus far.

CREB-binding protein (CBP) and p300 are structurally similar HATs that engage in H3K27Ac-mediated transcriptional co-activation and act as transcriptional scaffolds (He et al. 2021) (Fig. 2). CBP and p300 promote *insulin* and *Slc2a2* transcription via interactions with TFs NEUROD1, PDX1, and KLF11 (Zhang et al. 2021a). CBP and p300 also enhance β-cell gene expression through acetylation of the nucleosomes near cAMP-response element binding protein (CREB) target genes (Van de Velde et al. 2019). For instance, p300 regulates the expression of *TXNIP*, a key regulator of diabetic β-cell dysfunction, through an interaction with the ChREBP TF. Further investigations of the p300-TXNIP relationship in rat and human donor islets demonstrated that p300 expression levels increase in conditions of T2D, in parallel with TXNIP. Pharmacological and genetic inhibition of p300 reduces the expression of TXNIP under high glucose conditions, which increases insulin secretion and reduces β-cell apoptosis. These findings implicate p300 as a key mediator of TXNIP function and suggest that inhibition of its HAT activity may lessen TXNIP-mediated glucotoxicity and promote β-cell survival during diabetes (Bompada et al. 2016).

Conversely, p300 levels were reduced in β-cells and human islets exposed to T2D-like high glucose conditions (11–30 mM), islet amyloid polypeptide (IAPP), and inflammatory cytokines (Ruiz et al. 2018). This reduction was linked to glucolipotoxicity-induced proteasomal degradation of p300. Incongruent data on the role of p300 in diabetic conditions may reflect temporal differences in p300 regulation of pancreatic cells during development and disease.

In vivo analysis of *p300* deletion in developing endocrine progenitors, driven by *Ngn3-Cre*, revealed that mice did not develop glucose dysglycemia until after weaning (8 weeks old) (Zhang et al. 2021a). At this age, mice were glucose intolerant and had significantly reduced plasma insulin levels. Additionally, insulin content and β- and α-cell numbers were reduced relative to control mice. An islet-wide CBP deletion demonstrated a similar onset of glucose intolerance by 8 weeks of age, accompanied by reduced insulin content and β- and α-cell area. Reducing CBP expression in p300 knockouts (CBP^{Het};$p300^{KO}$) resulted in more severe glucose intolerance than either of the single gene knockouts ($p300^{KO}$ or CBP^{KO}) by 8 weeks of age; however, consistent reductions in insulin secretion, content, and β- and α-cell area were demonstrated. These reductions were also observed *in vitro* using a selective inhibitor of CBP/p300 HAT activity in rat islets (Wong et al. 2018).

Embryonic deletion of either p300 or CBP in endocrine progenitors reduced α-cell and β-cell area at E18.5, but the majority of tri-allelic knockout CBP^{Het};$p300^{KO}$ mice experience embryonic lethality or a complete lack of β- and α-cells (Wong et al. 2018). The observation of reduced β- and α-cell identity gene expression via RNA-seq of rat islets treated with this CBP/p300 inhibitor further supported this finding.

MOF (males absent on the first; also called MYST1 or KAT8) is a member of the MYST family (Moz, Ybf2/Sas2, Sas2, Tip60) of HATs and is the major enzyme that acetylates histone 4 at lysine 16 (H4K16). In the pancreas, MOF regulates DNA damage and α-cell differentiation (denDekker et al. 2020). Ablation of MOF expression in both lean and diet-induced obesity (DIO) mice decreased α-cell ratios (GCG^+/INS^+GCG^+). Fasting blood glucose levels and glucose tolerance were improved in MOF-deficient mice. These glucose homeostasis improvements were due to increased β-cell mass and enhanced glucagon-like peptide 1 (GLP-1) and insulin secretion (Guo et al. 2020). Furthermore, ChIP-Seq in an α-cell line treated with an MOF inhibitor, mg149, demonstrated that MOF regulates the expression of the TFs PAX6, NKX2.2, and FOXA2, essential for α-cell differentiation and function (Guo et al. 2021).

HDACs

Acetyl groups are removed by histone deacetylases (HDACs), which are also categorized into four classes based on their homology to yeast HDACs. Of these, class I and class II are the major HDACs involved in removing acetyl groups. HDACs are also grouped into two families based on whether they are zinc ion-dependent or nicotinamide adenine dinucleotide (NAD)-dependent. Class I HDACs are primarily localized within the nucleus and consist of the 1, 2, 3, and 8 isoforms. Class II HDACs are found in both the cytoplasm and nucleus and are categorized into IIa and IIb isoforms, depending on unique domain composition. Class IIa HDACs have an adaptor domain in the N-terminus and include isoforms 6 and 10. In contrast, class IIb HDACs have a tail domain at the C-terminus and include isoforms 4, 5, 7, and 9. Class III HDACs, also known as Sirtuin proteins, also localize to the cytoplasm and mitochondria. Class III HDACs are required for silencing transcription and are the only NAD^+-dependent family of HDACs. There is only one class IV HDAC, HDAC11, and it localizes to the nucleus and cytoplasm (Park and Kim 2020; Makkar et al. 2020).

CLASS I HDACs

All classes of HDACs are implicated in regulating β-cell function and metabolism. For instance, all 18 HDACs are expressed in β-cells, some class III HDACs are important for insulin secretion, class I HDACs regulate β-cell cytotoxicity, and class IIa HDACs modulate β- and δ-cell mass (Zhang et al. 2020a, 2019). HDAC1, a class I HDAC, has been extensively studied in the context of cell cycle progression in cancer, although recent analyses have expanded on this role in β-cells. Adenoviral-mediated overexpression of NKX6.1 in rat islets prompted the expression of HDAC1, which is sufficient to induce β-cell proliferation. Overexpression of HDAC1 demonstrated that β-cell replication is induced by the reduction of the cell-cycle inhibitor Cdkn1b and the upregulation of cell-cycle activators (Draney et al. 2018).

Prolonged treatment of rat islets with the HDAC inhibitors trichostatin A (TSA) or sodium butyrate (SB) increased insulin secretion. Investigations of the HDAC imparting this improvement demonstrated that HDAC1 inhibition increases glucose-stimulated insulin secretion (GSIS), reduces *ad lib* blood glucose levels, and improves glucose tolerance in chow-fed mice by increasing Tph1-mediated serotonin synthesis (Zhang et al. 2020a). HDAC1 is also implicated in regulating insulin signaling by modulating *IRS2*. IRS2 is upregulated in mouse Min6 β-cells that have increased insulin secretion due to a higher passage number. Treatment of Min6 cells with broad HDAC inhibitors like TSA and the HDAC1-specific inhibitor apicidin recapitulates the increase in *IRS2*. However, treatment with HDAC1 inhibitor alone is unable to increase insulin secretion in Min6 cells (Kawada et al. 2017).

Deletion of *HDAC3* in adult mouse β-cells using a tamoxifen-inducible Cre driven by the mouse *insulin 1* promoter (*MIP-Cre*^{ERT}) similarly improves glucose tolerance and insulin secretion (Remsberg et al. 2017). Cistromic analyses revealed enrichment for HDAC3 binding near genes involved in endocrine cell development and the upregulation of GSIS targets, like *Rsg16* and *S100a6* (Calcyclin). Constitutive deletion of HDAC3 in mouse β-cells with a Cre recombinase driven by the rat *insulin 2* promoter (*RIP-Cre*) demonstrated contrasting results, with reduced insulin secretion and significant impairment of glucose tolerance upon HDAC3 loss (Chen et al. 2016). Aside from the limitations of differing Cre models, these varying results suggest HDAC3 may regulate β-cell identity and maintenance through distinct pathways.

CLASS II HDACs

Class II HDAC proteins, consisting of HDAC7, HDAC4, and HDAC6, also regulate insulin secretion. Islets from individuals with T2D display higher HDAC7 expression than islets from non-diabetics. Mitochondrial respiration is reduced in rat β-cells overexpressing HDAC7, indicating that dysregulation of mitochondrial function may contribute to T2D-associated insulin secretion and content reduction (Daneshpajooh et al. 2017). Studies of three individuals with T2D bearing missense mutations in *HDAC4* (H227R, D234N, and E374K), and recapitulation of these mutations in mouse β-cells, revealed impaired insulin secretion and reduced expression of the β-cell TFs PDX1, NEUROD1, and FOXO1 (Gong et al. 2019). Investigators found that HDAC4 mutations may impart these defects through disrupted FOXO1 nuclear translocation resulting from a lack of FOXO1 acetylation. Additionally, HDAC6 regulates insulin signaling, and its inhibition in mouse β-cells and islets using tubacin treatment resulted in the reduction of P-AKT and P-S6 signaling effectors. Furthermore, *Lepr*$^{db/db}$ mice, which have impaired insulin signaling relative to lean mice, have reduced HDAC6 protein levels (Inoue et al. 2021). However, a study of class IIa HDACS in INS1E rat β-cells, cultured in normal and high glucose, suggests that HDAC4 may be dispensable for β-cell function; under- and overexpression of HDAC4 did not alter insulin expression, secretion, or β-cell viability (McCann et al. 2019).

CLASS III HDACs

Sirtuins are class III HDACs with NAD^+-dependent deacetylation activity. Of the seven sirtuins, SIRT1 and SIRT6 have been extensively studied and shown to regulate metabolism (Zhou et al. 2018). SIRT6 carries out deacetylation of H3K18Ac, H3K9Ac, H3K56Ac, and TFs like FOXO1 (Fig. 2). Reduction of SIRT6 expression in Min6 mouse β-cells impairs GSIS, with a similar dysregulation occurring in β-cell-specific and pancreas-wide SIRT6 knockout mice. Loss of SIRT6 in either β-cells or PDX1-expressing pancreatic cells results in glucose intolerance (Xiong et al. 2016a). This dysregulation of GSIS and glucose tolerance may result from disruption of glucose oxidation, mitochondrial damage, and disruption of calcium flux. Further studies of SIRT6 β-cell-specific knockouts suggest that GSIS dysregulation may be the result of disrupting the FOXO1 glucose sensing pathway (Song et al. 2016). FOXO1 negatively regulates PDX1, which promotes GLUT2 (*Slc2a2*) expression. In SIRT6 β-cell-specific knockout mice, FOXO1 is upregulated, and in turn, both PDX1 and GLUT2 are downregulated at the mRNA and protein levels. These findings, along with observations of SIRT6 deacetylation of FOXO1, suggest that SIRT6 regulates glucose sensing through interactions with PDX1 and GLUT2.

In addition to its roles in regulating GSIS, SIRT6 also protects β-cells against lipotoxicity. SIRT6 expression is reduced in islets of DIO mice, aged mice, and Min6 cells and mouse islets treated with palmitate (Xiong et al. 2016b). Overexpression of SIRT6, however, rescues Min6 mouse β-cells from palmitate-induced apoptosis. SIRT6 also protects β-cells by downregulating *TXNIP* expression (Qin et al. 2018). Other Sirtuins, including SIRT1–4, offer similar protection against β-cell apoptosis and insulin secretion aberrations. Overexpression of SIRT1–3 increases GSIS, while ablation of SIRT4 decreases GSIS (Zhang et al. 2016). In addition to the aforementioned β-cell roles, SIRT1 also regulates α-cell functional identity. Treatment of diabetic mice with a SIRT1 activator prevents diabetes-associated α-cell hyperplasia and glucagon increase, whereas treatment of the In-R1-G9 α-cell line with a SIRT1 activator significantly reduces glucagon secretion, α-cell proliferation, and *Arx* expression (Zhang et al. 2018).

Histone Acetylation Readers

Epigenetic readers must recognize acetylation marks on lysine tails of histones to modulate gene transcription. Bromodomain and extraterminal (BET) proteins specifically read acetylated lysines and enable acetylation-associated chromatin opening. While mainly studied in the context of cancer, recent studies have appreciated roles for BET proteins in adipogenesis and pancreas development (Boyson et al. 2021; Lee et al. 2017). BET proteins, encoded by *Brd2*, *Brd3*, and *Brd4*, are expressed in mouse pancreatic buds beginning at E11.5 and are maintained in tissue explants up to E16.5. Pharmacological inhibition of BET proteins in mouse pancreatic bud explants and human iPSC (induced pluripotent stem cell) pancreatic progenitors increased *Ngn3*, *somatostatin*, *glucagon*, and *ghrelin* expression (Huijbregts et al. 2019). *MafA* and *Nkx6.1* increased following BET inhibition; however, insulin expression and content were only upregulated following the removal of chronic BET inhibition. *In vitro* inhibition of BET proteins potentiates endocrine cell marker expression, suggesting that these factors modulate endocrine cell development and differentiation.

2.2.2 Histone Methylation

Histone methylation silences or activates gene expression through the placement of methyl groups on arginine (R) and lysine (K) residues (Fig. 2). Methylation is carried out by histone methyltransferases (HMTs) and can be removed by histone demethylases (HDMs). The different methylation states—mono-, di-, tri-, or poly-methylation—and their specific position determine whether gene transcription will be repressed or activated. For example, methylation of H3K79, H3K36, and H3K4 is generally associated with gene activation, whereas methylation of H4K20, H3K27, and H3K9 is associated with silencing (Hyun et al. 2017). Variability in histone methylation patterns facilitates complex regulation of pancreas development and islet cell differentiation.

Polycomb Group Proteins

Polycomb group proteins are transcriptional regulators that form epigenetic complexes, including the two most extensively studied Polycomb Repressive Complex 1 (PRC1) and PRC2 (Dumasia and Pethe 2020). In addition to recruiting PRC1 activity through methylation, PRC2 mono-, di-, and tri-methylates H3K27, which regulates endocrine pancreas and β-cell differentiation. EZH1 and EZH2 are the PRC2 methyltransferases that carry out this regulation. Additional complex components EED, SUZ12, and RBAP48 contribute to gene silencing and are essential for regulating endocrine pancreas specification, β-cell proliferation, and maintenance of β-cell identity (Lu et al. 2018; Dahlby et al. 2020). Most studies of the PRC2 methyltransferases focus on EZH2, primarily due to its higher methyltransferase activity and the fact that it is expressed in cells with greater rates of division (Golson 2021). EZH2 engages in transcriptional repression and activation, either alone or via interactions with polycomb components. Ablation of Raptor, an essential regulatory protein of mTORC1, in NGN3$^+$ progenitors demonstrated that EZH2 is involved in the G2/M phase and proliferation of neonatal β-cells (Wang et al. 2019). This observation is further supported by findings that β-cell-specific EZH2 knockout mice experience mild diabetes due to disruption of β-cell proliferation (Lu et al. 2018). Additionally, EZH2 protects β-cells from glucose- and lipotoxicity-induced ER stress and apoptosis (Dahlby et al. 2020).

The role of EED in the complex is to provide structural support to PRC2 methyltransferases. Upon EED loss, both EZH1 and EZH2 lose HMT activity. Despite this lack of HMT activity, β-cell-specific mouse EED deletion models have no disruption of islet morphology or development before 8 weeks of age. This observation suggests that another PRC2 complex member has compensatory function and that EED is dispensable for PRC2-mediated regulation of β-cell proliferation. As these mice age past 8 weeks, they experience glucose intolerance, impaired insulin secretion, and reduced expression of β-cell maturity markers PDX1 and Urocortin-3 (UCN3), suggesting that EED is important during adulthood for β-cell maintenance (Lu et al. 2018).

Demethylases

H3K27me facilitated by PRC2 is edited by HDMs, including the Jumonji domain-containing demethylases (JMJCs), JMJD3, and UTX. Dysregulation of H3K27me HDMs prompts disruption of endocrine cell-fate specification. Mouse pancreatic cells treated with GSK-J4, an inhibitor of H3K27 demethylation, and thereby JMJD3 and Utx activity, exhibit a reduction in NGN3 *trans*-activation activity, demonstrated by reduced expression of its target genes *Pax4*, *NeuroD1*, and *Nkx2.2* (Fontcuberta-PiSunyer et al. 1861). Additionally, JMJD3 regulates endocrine differentiation by mediating the transition from $NGN3^{Lo}$ to $NGN3^{Hi}$ expression, which is the transition of progenitors into endocrine-committed cells (Yu et al. 2018). Deleting JMJD3 in PDX1-expressing pancreatic cells decreases the ratio of $NGN3^{Hi}$ cells at E13.5. By 6 weeks, these mice have impaired glucose tolerance and reduced insulin secretion. These findings suggest that JMJD3 demethylase activity is important for pancreas and islet development.

Jumonji And AT-Rich Interaction Domain Containing 2 (JARID2) is an atypical member of the JMJC family that regulates H3K27me. It does not engage in demethylation but instead acts as an accessory subunit involved in the recruitment of PRC2 to chromatin (Hojfeldt et al. 2019). Deletion of JARID2 in PDX1-expressing pancreatic progenitors reduces β-cell mass, causes glucose intolerance, and impairs the expression of β-cell maturity genes, including *MafA*, *Pcsk1*, *G6pc2*, and *Iapp* (Cervantes et al. 2017).

H3K27me3 has also been implicated in regulating β-cell function by mediating inflammatory cytokine levels. The histone demethylase KDM6B is expressed in macrophages and increases the transcription of proinflammatory cytokine genes by removing promoter and enhancer H3K27me di- or tri-methyl marks. Treatment of human islets and Min6 β-cells with the GSK-J4 inhibitor protects against cytokine-induced apoptosis. Additionally, preservation of H3K27me3 by treating INS-1 rat β-cells with GSK-J4 increases insulin content and secretion, as well as *Ins1* and *Ins2* mRNA levels, relative to untreated cells. GSK-J4 treatment also significantly reduces the expression of the ER stress markers CHOP and DP5 (Backe et al. 2018). Despite the observed protection against cytokine-induced apoptosis, *in vivo* administration of GSK-J4 to STZ-induced diabetic mice does not rescue hyperglycemia, even though plasma insulin levels increase (Backe et al. 2019).

TrxG Complexes

Trithorax group (TrxG) complexes impact gene expression through H3K4me and consist of HMTs, SET1A/B, and MLL1–9, the core component proteins DPY30, RBBP5, WDR5, and ASH2L (Schuettengruber et al. 2017). DPY30 is required for TrxG catalytic activity and recruitment of TrxG complexes to chromatin. Deleting DPY30 in PDX1-expressing pancreas progenitors reduces H3K4me3, decreases the proliferation of tip and trunk progenitor cells, increases progenitor apoptosis, and dysregulates acinar cell terminal differentiation (Campbell et al. 2019). Additionally, studies of DPY30 via deletion from NGN3-expressing endocrine progenitors prompted hyperglycemia, glucose intolerance, reduced insulin content, and reduced expression of β-cell maturity and GSIS genes in 3-week-old mice (Campbell et al. 2021). WDR5 provides scaffolding support to TrxG complexes and also acts as an H3K4me3 reader (Guarnaccia and Tansey 2018). In mouse pancreatic progenitor spheroids, loss of WDR5 expression results in reduced expression of endocrine and acinar genes and increased ductal gene expression, including *Car2*, *Krt19*, and *Muc20*. These findings suggest that WDR5 is important for regulating acinar and endocrine gene expression during pancreas development (Campbell et al. 2019). Similar to PRC2, TrxG components interact with other proteins to regulate gene methylation. Ubiquitin ligase CRL4 ubiquitinates WDR5 and reduces H3K4me3 in neuronal cells; however, this relationship and its impact on pancreas development has yet to be studied (Nakagawa and Xiong 2011).

Arginine Methylation and PRMTs

Arginine methylation of histones is another regulator of gene expression facilitated by protein arginine methyltransferases (PRMTs). There are three different classes of PRMTs based on their enzymatic activity and the three forms of methylarginines they produce: mono-methylarginines formed by class three PRMTs, symmetric dimethyl arginine formed by class two PRMTs, and asymmetric dimethyl arginine formed by class one PRMTs. Dimethyl arginine symmetry is defined by the placement of methyl groups on terminal nitrogen atoms (Bedford and Clarke 2009; Blanc and Richard 2017).

PRMT1 is a class I PRMT that methylates histone 4 (H4R3me2a) and is implicated in the regulation of pancreas development and β-cell identity. Ablation of PRMT1 in mouse $PDX1^+$ pancreas progenitors results in lower body weight and hyperglycemia at 4 weeks and reduced β-cell mass and hypoplastic pancreata at P7

(Lee et al. 2019). These mice also have defects during the secondary transition, wherein progenitors fail to exhibit the typical increase in proliferation and differentiation. Additionally, PRMT1 knockout mice present with prolonged NGN3 expression, highlighting a dysregulation in endocrine cell differentiation. Subsequent degradation assays confirm that PRMT1 is required for the characteristic reduction of NGN3 protein following E15.5.

H4R3me2a ChIP-Seq and ATAC-Seq analyses in the Min6 β-cell line demonstrated that PRMT1 and H4R3me2a maintain open chromatin architecture at β-cell identity genes also regulated by TFs, including MafA, NeuroD1, and FoxA1 (Kim et al. 2020). Deletion of PRMT1 during β-cell development using *RIP2-Cre* disrupts islet formation. By 12 weeks of age, these mice had reduced β-cell maturity gene expression, reduced insulin granule volume and density, were glucose intolerant, and had GSIS defects. Mice with an inducible deletion of PRMT1 in PDX1-expressing β-cells demonstrated a similar glucose intolerance and disruption of GSIS as mice with a *RIP2-Cre*-driven PRMT1 deletion by 12 weeks; they also displayed decreased expression of MAFA and GLUT2. Lineage-tracing analyses of PRMT1-deficient β-cells unveiled polyhormonal cells and loss of typical β-cell identity, with reduced expression of insulin and UCN3.

PRMT4 is another class I PRMT that regulates β-cell function, specifically insulin secretion. Treatment of INS-1 and HIT-T15 β-cell lines with high glucose prompted the upregulation of PRMT4 and the corresponding H4R3me2a mark (Kim et al. 2015). Pharmacological or genetic inhibition of PRMT4 activity reduced GSIS, suggesting that its asymmetric demethylation activity is important for insulin secretion.

PRMT5 is a class II PRMT that carries out symmetric methylation of histone H2A (H2AR3me2s), H3 (H3R8me2s), and H4 (H4R3me2s). Inducible deletion of PRMT5 in Pdx1-expressing adult β-cells resulted in glucose intolerance, reduced GSIS, reduced insulin expression, and elevated *ad lib* blood glucose (Ma et al. 2020). Additionally, in the rat INS-1 β-cell line, PRMT5 and H3R8me2s were enriched at the *insulin* promoter. Altogether, these data suggest that PRMT5 regulates *insulin* transcription and multiple insulin-secretion-related genes and processes; for example, GLP1R, the incretin receptor, which amplifies GSIS, is downregulated in response to decreased PRMT5 expression.

Menin is a scaffolding protein that interacts with epigenetic modifiers, like PRMT5, to facilitate gene transcription and has previously been implicated in regulating β-cell proliferation. Studies of its relationship with PRMT5 in β-cells, using a Min6 cell ChIP, demonstrated that both PRMT5 and Menin bind to the *Glp1r* promoter (Muhammad et al. 2017). siRNA-mediated PRMT5 knockdown significantly increased *Glp1r* expression, suggesting that Menin recruits PRMT5 to suppress *Glp1r*. Downstream of GLP1R, PRMT5 was implicated in regulating the phosphorylation of FOXO1 and CREB, as siRNA knockdown of PRMT5 increased their phosphorylation. These findings further suggest a role for PRMT5 in regulating β-cell function and survival.

2.2.3 Histone Ubiquitination

Histone ubiquitination is an epigenetic modification that informs gene transcription through the labeling of lysine or methionine residues with ubiquitin marks (Fig. 2). The number of ubiquitin marks deposited dictates varying gene expression outcomes, with polyubiquitination primarily associated with proteasomal degradation and monoubiquitination linked to transcriptional regulation. Ubiquitination is additionally involved in a multitude of signaling processes, including DNA-damage response and cell-cycle control (Damgaard 2021). E3 ubiquitin ligases are responsible for depositing the marks under the guidance of E1 ubiquitin-activating and E2 ubiquitin-conjugating enzymes. Alternatively, ubiquitin marks can be removed by deubiquitylases (DUBs). Both ubiquitin ligases and DUBs are important for mediating control of gene transcription and are implicated in numerous disorders, including diabetes.

PRC1 and PRC2 engage in ubiquitination and methylation of the *Arf/Ink4a* locus. PRC1 comprises the ubiquitin ligases RING1B/RING2/RNF2 and BMI1. Both RING1B and BMI1 are recruited by PRC2 methylation and, subsequently, ubiquitinate histone H2A at lysine 119 (H2AK119) (Dhawan et al. 2016). Together, the epigenetic modifications mediated by PRC1 and PRC2 work to repress *Arf/Ink4a* to promote β-cell proliferation. As islets age, Bmi1 stability decreases, prompting the reduction of PRC1-mediated ubiquitination activity and causing a decrease in β-cell replication. In BON cells, a human pancreatic neuroendocrine tumor cell line, the expression of *Arf/Ink4a* and BMI1 is induced by NGN3, which may facilitate both endocrine pancreas differentiation and hormone secretion in mature pancreata (Solorzano-Vargas et al. 2019).

Ring finger 20 (RNF20) and 40 (RNF40) are ubiquitin ligases that monoubiquitinate histone H2B at lysine 120 (H2Bub1) as either a homodimeric or heterodimeric complex. In β-cells, these ligases interact with the transcription factor ISLET-1, which is vital for regulating β-cell development, identity, and function (Ediger et al. 2014; Du et al. 2009). *In vitro*, RNF20/40 mediate the expression of crucial β-cell genes, like *Glut2* and *MafA*, through occupancy of their regulatory domains, potentially based on recruitment by ISLET-1 (Wade et al. 2019). Additionally, independent siRNA-mediated knockdown of *Rnf20* and *Rnf40* reduced GSIS in Min6 cells. Loss of *Rnf20* and *Rnf40* in Min6 cells also reduced the expression of *Ucp2,* which prompted an increase in mitochondrial ROS that may contribute to the dysregulation of insulin secretion (Wade et al. 2019).

Histone ubiquitination also regulates DNA damage within β-cells, for instance, hyperglycemia-inducing impairments prompted by aberrant expression of GSIS factors such as GLUT2 (Zhong et al. 2018). Ubiquitination of H2A and H2B remedies this damage through the DNA damage response (DDR). The PRC1/2 and RNF20/40 ubiquitin ligases have each been implicated in mediating DDR by labeling double-stranded breaks (DSBs), potentially to initiate repair signaling (Smeenk and Mailand 2016). Additionally, ubiquitin ligases RNF168 and RNF8 carry out DSB-responsive signaling to mediate DDR in interactions with BRCA1 (Uckelmann and Sixma 2017). Overall, histone ubiquitination is involved in regulating β-cells throughout their entire life cycle, starting with the regulation of proliferation and response to DNA damage-associated apoptosis.

2.2.4 Phosphorylation, Kinases, and Phosphorylases

Histone phosphorylation can regulate gene transcription, condense chromatin, repair DNA damage, and can occur on all histones at tyrosine, serine, and threonine residues. The addition of phosphate marks on histone tails is carried out by kinases and removed by phosphorylases (Rossetto et al. 2012). Thus far, studies of histone phosphorylation in islet and β-cell development are limited. However, histone phosphorylation is involved in DNA-damage repair (Kuo and Yang 2008; Millan-Zambrano et al. 2018), which is implicated in diabetes pathogenesis (Shimizu et al. 2014; Horwitz et al. 2018; Tay et al. 2019). This relationship suggests that histone phosphorylation may regulate islet and/or β-cell stress response, but the mechanisms have yet to be defined.

3 Non-Coding RNAs

Non-coding RNAs (ncRNAs) are RNAs that are not translated into protein. Although previously considered nonfunctional, ncRNAs are now known to have critical roles in regulating gene expression. There are multiple types of ncRNAs, including long ncRNAs (lncRNAs), which are longer than 200 nucleotides, piwi-interacting RNA (piRNAs), which are between 26 and 31 nucleotides, and microRNAs (miRNAs), which are between 19 and 24 nucleotides. All these ncRNAs are single-stranded; however, small interfering RNAs (siRNAs) are double-stranded (Li and Liu 2019; Wei et al. 2017).

3.1 lncRNAs

lncRNAs can regulate gene expression by modifying chromatin states and regulating transcriptional or translational complexes (Li and Liu 2019). As genome-wide studies of regulators of pancreas and β-cell development become more prevalent, lncRNAs are proving to be essential mediators.

3.1.1 βLINC1

βLINC1 (β-cell Long Intergenic Noncoding RNA 1) is an lncRNA localized upstream of *Nkx2.2* and is expressed throughout the pancreas and islets. Ablation of *βLINC1* expression in the Min6 mouse β-cell line and global knockout mice demonstrated a significant reduction of *Nkx2.2*, suggesting *βLINC1* positively regulates Nkx2.2 expression (Arnes et al. 2016). Adult *βLINC1* knockout mice experience mild glucose intolerance, increased insulin secretion in low glucose conditions, reduced β-cells, and an increase in δ-cell number at 16 weeks. This dysregulation of glucose homeostasis in *βLINC1* knockout mice results from endocrine cell development disruption, demonstrated by reductions in β-cells and upregulation of δ-cell number at E15.5. Transcriptomic analysis of pancreata from E15.5 *βLINC1* knockout mice showed a significant reduction of TF-encoding mRNAs involved in endocrine progenitors and β-cell functional identity, including *Ngn3*, *Sox9*, *Nkx2.2*, *Neurod1*, and *Nkx6.1*. Positional mapping of the dysregulated genes demonstrated that the *βLINC1* transcript predominately regulates the expression of neighboring genes on chromosome 2.

3.1.2 MEG3

MEG3 (Maternally Expressed Gene 3) is an lncRNA whose expression is reduced in islets from donors with T2D (Kameswaran et al. 2014) and mouse models of both type 1 (nonobese diabetic, NOD) and type 2 diabetes (*Lepr*$^{db/db}$) (Wang et al. 2018), suggesting it is an important regulator of β-cell function. In human islets, an assay that captures regulatory RNA-mRNA pairings indicated that *MEG3* binds *islet amyloid polypeptide* (*Iapp*) and *tumor protein 53 inducible nuclear protein 1* (*TP53INP1*), both of which promote apoptosis (Kameswaran et al. 2014). In mouse islets, *MEG3* promotes transcription of *MafA* via the inactivation of *Rad21, Smc3, and Sin3α,* whose gene products repress *MafA* transcription (Wang et al. 2018). siRNA-mediated knockdown of *Rad21*, *Smc3*, and *Sin3α* increased *MafA* expression and GSIS in Min6 β-cells and mouse islets. Additionally, *MEG3* negatively regulates *Rad21*, *Smc3*, and *Sin3α* via EZH2 action at the target promoters. These results suggest that *MEG3* promotes *MafA* transcription by inducing the repression of its negative regulators. Further ChIP-qPCR experiments support this notion by revealing that *MEG3* knockdown prompted increased occupancy of the *MafA* promoter by these negative regulators (Wang et al. 2018).

3.1.3 PLUTO

PDX1 Locus Upstream Transcript *(PLUTO)* is an lncRNA that regulates islet function by mediating TF expression. *PLUTO* is localized upstream of *Pdx1* in an enhancer cluster, and knockdown in the EndoC-βH1 human β-cell line leads to reduced contact between *PDX1* and its upstream enhancers. Reduction of *PLUTO* transcription significantly inhibits *Pdx1* expression through regulation of chromatin structure, thereby acting as a key mediator of β-cell function and identity. It is reduced in environments of β-cell dysfunction, including T2D and glucose intolerance (Akerman et al. 2017).

3.1.4 Paupar

Paupar (Pax6 upstream antisense RNA) is a nuclear, α-cell-enriched lncRNA that regulates α-cell function through interaction with *Pax6*. PAX6 comprises two major isoforms, PAX6 and PAX6 5a. The isoform PAX6 5a is enriched in α-cells and is required to activate α-cell target genes. *Paupar* is localized to α-cell nuclei, where it regulates isoforms of PAX6 via alternative splicing and the expression of PAX6 targets including *Gcg*, *MafB*, *Arx*, and *NeuroD1*. Similar to *MEG3* and *PLUTO*, expression of *Paupar* is reduced in *Lepr*$^{db/db}$ mice. Physiological analysis of 6-week-old *Paupar* knockout mice demonstrated that they are hypoglycemic, secrete less glucagon than wild-type control littermates, and have reduced α-cell area (Singer et al. 2019).

3.1.5 ROIT

Regulator of insulin transcription (ROIT) is an lncRNA enriched in pancreatic islets. ROIT expression is reduced in the islets of obese humans and mouse models of obesity, including high-fat diet (HFD)-fed, *Lepr*$^{db/db}$, and *Lep*$^{ob/ob}$ mice, as well as upon exposure to high lipid concentrations (Zhang et al. 2020b). The reduction of *ROIT* in obesity models is partly regulated by its interaction with the TF HNF1β, which has a binding site within the *ROIT* promoter and has a similarly reduced expression in obese contexts. Moreover, *HNF1β* overexpression can rescue *ROIT* levels in palmitate-treated mouse β-cells. ROIT further regulates β-cell function by driving insulin synthesis and *Nkx6.1* expression. Overexpression of *ROIT* increases insulin content and gene transcription, whereas suppression or ablation inhibits synthesis. ROIT elevates *Nkx6.1* expression, perhaps by preventing methylation of its coding region. Degradation analyses in Min6 β-cells demonstrated that *ROIT* overexpression increases the ubiquitination of DNMT3A, which is involved in β-cell identity and function (Parveen and Dhawan 2021). These findings, alongside the rescue of *Nkx6.1* expression via DNMT inhibitors in *ROIT* knockdown cells, suggest that ROIT epigenetically and transcriptionally maintains *insulin* expression.

3.1.6 MALAT1

MALAT1 is an lncRNA extensively studied in cancer biology. Initial islet studies demonstrated that *MALAT1* interacts with microRNA-17 (miR-17) as a competing endogenous RNA (ceRNA) to regulate β-cell function upon exposure to stressors. ceRNAs are RNA molecules that share target mRNAs with miRNAs known as microRNA recognition elements (mREs). ceRNAs regulate gene expression by competitive binding to mREs (Kartha and Subramanian 2014) (Fig. 3). While miR-17 inhibits *TXNIP* expression, upon incidents of stress, such as incubation of Min6 β-cells with cigarette smoke extract, *MALAT1* sequesters miR-17 activity and allows for *TXNIP* upregulation. *MALAT1* overexpression

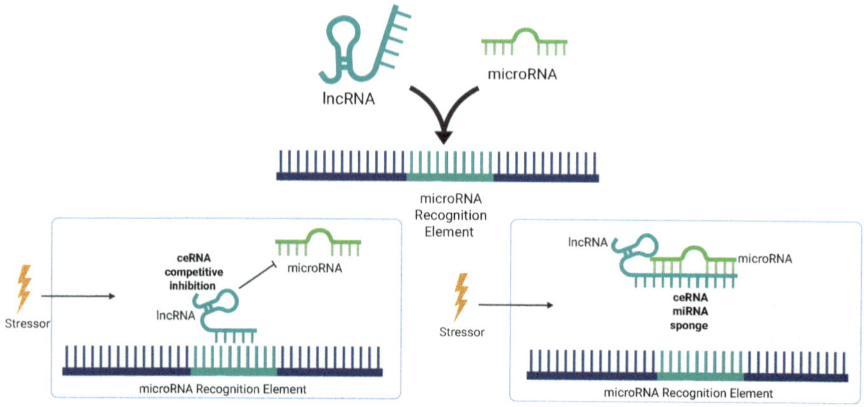

Fig. 3 lncRNA as competitive inhibitors. Molecules that share target miRNAs or miRNA recognition elements (mREs) are known as competing endogenous RNAs (ceRNAs). lncRNAs and miRNAs can undergo competitive inhibition under stress (left), in a process in which the lncRNA binds the target sequence. Alternatively, the lncRNA can bind the miRNA itself (right), thus acting as a ceRNA miRNA sponge. Created with BioRender.com

contributes to β-cell dysregulation by facilitating the cigarette smoke extract-induced reduction in GSIS and *MafA* expression, as siRNA-mediated *MALAT1* knockdown restored both to levels comparable to control β-cells (Sun et al. 2018). As hyperglycemia develops in NOD mice, *MALAT1* expression increases in an age-dependent manner, and concomitantly, *Pdx1* expression is significantly reduced. *MALAT1* reduces *Pdx1* expression by decreasing histone acetylation at its promoter. In NOD islets, H4 acetylation of the *Pdx1* promoter is reduced, and similarly, overexpression of *MALAT1* reduces H3 acetylation in mouse islets and β-cells. *MALAT1*-mediated reduction of *Pdx1* expression results in reduced insulin secretion (Ding et al. 2020).

3.1.7 TUNAR

The lncRNA *TUNAR* is expressed in pancreatic islets and enriched specifically in β-cells. Overexpression of *TUNAR* in the EndoC-βH1 human β-cell line stimulates proliferation, mediated by canonical Wnt signaling. Interestingly, *TUNAR* is upregulated by the inhibition of GSK3, an action that stabilizes β-catenin (Zhou et al. 2021). Additionally, TUNAR negatively regulates the WNT agonist DKK3, which is significantly upregulated in T2D β-cells. siRNA-mediated knockdown of EZH2 in EndoC-βH1 cells significantly increases DKK3 levels, and *TUNAR* was significantly enriched in EZH2 RNA immunoprecipitation pulldowns. These findings suggest that *TUNAR* may repress DKK3 expression through interactions with EZH2.

3.1.8 βFaar

β-cell function and apoptosis regulator (*βFaar*) is another lncRNA enriched in pancreatic islets. *βFaar* expression is reduced in obese mouse islets and upon exposure to palmitate and inflammatory cytokines (Zhang et al. 2021b). These effects are thought to be driven by DNMT3a and DNMT3b methylation of the *βFaar* promoter. Ablation of *βFaar* in mouse islets and β-cells reduces insulin secretion, insulin content, and *Insulin, Creb1,* and *Neurod1* expression, prompting glucose intolerance. Additionally, *βFaar* silencing increases β-cell apoptosis, BAX protein levels, and Caspase 3 cleavage and reduces BCL2 protein and pro-caspase-3 levels.

3.2 miRNAs

miRNAs are ncRNAs that post-transcriptionally silence gene expression. Immature miRNAs (pre-miRNAs) are hairpin structures that are processed by DICER into small RNA duplexes, one of which is incorporated into a miRNA-induced silencing complex (miRISCs) with an Argonaute protein to silence target mRNAs (O'Brien et al. 2018). miRNAs have been implicated in regulating cell differentiation, proliferation, development, and metabolism across many tissue types. In the pancreas, miRNAs are involved in the specification of endocrine progenitor differentiation, regulation of islet cell maturation, disallowed gene expression, and maintenance of β-cell identity (Martinez-Sanchez et al. 2016). Additionally, many miRNAs have been associated with diabetes. For example, miR-132 is upregulated in T2D animal models and may regulate insulin secretion. Pharmaceutical antagonism of miR-132 in mice results in reduced blood glucose levels, increased plasma insulin, and improved insulin secretion. Similar results were also observed in human islets treated with miR-132 antagonists; however, miR-132 did not exhibit differential expression in individuals with T2D. RNA-Seq experiments using human T2D islets demonstrated dysregulation of miR-132 targets, including *FOXA1*, *SLC25A28,* and *DESI2,* suggesting a link between the rodent and human observations (Bijkerk et al. 2019).

miR-132 also has impacts on β-cell proliferation. miR-132 and other miRNAs were upregulated in a partially pancreatectomized mouse model (Mziaut et al. 2020). Overexpression of miR-132 in EndoC-βH1 cells increased proliferation, as demonstrated by BrdU incorporation. Conversely, deletion of miR-132 in a mouse β-cell line reduced BrdU uptake and increased expression of the apoptotic markers cleaved caspase-3 and caspase-9. miR-132 regulates proliferation via binding interactions with *Pten*, *Akt*, and *FoxO3*. Overexpression of miR-132 downregulates *Pten* expression and upregulates phosphorylation of AKT and FOXO3 proteins, demonstrating a role in regulating cell proliferation and apoptosis mediators.

miRNAs also interact with Sirtuins to regulate oxidative stress, β-cell apoptosis, and insulin secretion. miR-34a negatively regulates SIRT1. Knockdown of miR-34a expression in rat β-cells treated with high glucose and palmitate increased expression of SIRT1 and reduced apoptosis and oxidative stress. The lncRNA *LEGLTBC* is a miR-34a ceRNA that promotes SIRT1 expression to protect β-cells against glucolipotoxicity (Kong et al. 2019). miR-212-5p is a miRNA that negatively regulates another Sirtuin, SIRT2. Overexpression of miR-212-5p reduces *SIRT2* expression in mouse β-cells and impairs glucose-stimulated insulin secretion. Within the classically activated and proinflammatory M1 macrophage exosomes, miR-212-5p levels were elevated and contributed to disrupted insulin secretion (Qian et al. 2021).

4 Environment, Epigenetics, and Diabetes

Comprehensive Genome-Wide Association Studies (GWAS) have demonstrated differential DNA methylation patterns in human T2D islets compared to non-diabetic islets (Volkmar et al. 2012; Volkov et al. 2017). Although varying methylation states have been linked to the onset of diabetes, what causes these modifications and whether they are inherited or acquired throughout life remain unanswered. As the development of diabetes has been linked to both genetic and environmental factors, it is likely that environmental triggers contribute to epigenetic alterations.

Exposure to nutritional and environmental factors during development has been proposed as a potential source for epigenetic changes. Nutrients are hypothesized to induce epigenetic changes as a way for islet cells to respond or compensate for energy demands. Known as "developmental programming," environmental factors during critical times in fetal gestation may have lifelong consequences on human health. Both overnutrition and undernutrition can negatively affect islet function. Exposure to specific macromolecules has been linked to epigenetic changes in islets, and fetal nutrient abundance regulates the size of endocrine progenitor pools and β-cell mass. Palmitate, a long-chain saturated fatty acid, induces changes in CpG methylation that impact islet genes involved in glucose and lipid metabolism, thereby decreasing insulin secretion (Hall et al. 2014). High levels of leucine inhibit β-cell proliferation in mouse pancreatic bud cultures (Rachdi et al. 2012). In humans, deficiency in certain nutrients like vitamin B12 and folate during pregnancy is predictive of greater adiposity and risk for developing T2D (Yajnik et al. 2008).

In rodent models, postnatal overnutrition results in differential methylation patterns in islets at weaning (Li et al. 2019). Offspring of mice fed an HFD before and during pregnancy develop glucose intolerance, insulin resistance, and β-cell proliferation defects relative to the offspring of mice initially fed an HFD, but moved to a standard chow diet 9 weeks before and during pregnancy (Liu et al. 2020). Decreased nutrition in the fetal state also impacts β-cell development and maturation. In utero exposure to 60% caloric restriction results in a significant reduction in the $PDX1^+$ endocrine cell proliferation at E13 relative to control mouse embryos (Elghazi et al. 2017). Similarly, in utero exposure to a low-protein diet predisposes mice to glucose intolerance and insulin resistance later in life (Alejandro et al. 2020).

5 Epigenetics as a Tool for Islet and β-Cell Reprogramming

While the above discussion has focused on the endogenous roles of epigenetics, the importance of these mechanisms may also be exploited for therapeutic purposes. The key role of chromatin accessibility in the differentiation process makes modulation of epigenetic targets a promising avenue for driving β- and islet-cell reprogramming for replacement therapies.

Since many miRNAs have been characterized throughout pancreatic endocrine cell development and in adult islet cell function, their use in iPSC differentiation toward β-cells is being explored. Differential expression of 18 miRNAs is detected during the transition of human iPSCs into insulin-producing cells, including miR-9-5p, miR-124a, and miR-375, which were previously implicated in endocrine pancreas development (Sebastiani et al. 2017). These miRNAs that were differentially expressed during differentiation regulate the key TFs MAFB and FOXA2, highlighting their necessity for the transition to functional β-cells. Others have observed that direct repression of SMAD7 and TGIF2 by miRNA-181c-5p promotes the differentiation of hiPSCs into insulin-producing cells (Li et al. 2020). Overexpression of this miRNA increased PDX1, NKX6.1, and MAFA during differentiation, and transplantation of miR-181c-5p-overexpressing iPSCs into diabetic mice improved blood glucose relative to control iPSCs. These data suggest that treatment with epigenetic factors can increase the efficiency of using patient-specific hiPCs for β-cell replacement.

Another epigenetic factor being explored in β-cell reprogramming is the histone demethylase LSD1, which inhibits enhancer chromatin activity by catalyzing the removal of methyl marks from H3K4. LSD1 inactivates enhancers of pluripotency genes in stem cells upon differentiation and is required during endocrine progenitor differentiation (Whyte et al. 2012; Vinckier et al. 2020). Inhibition of LSD1 with tranylcypromine promotes the differentiation of hiPSCs into insulin-producing cells (Vinckier et al. 2020; Yang et al. 2020). LSD1-inhibited cells secreted significantly more insulin post-transplantation than scramble controls. This experiment highlights the ability to use epigenetic factors *in vivo* and *in vitro* to manipulate chromatin states and cell identity.

6 Conclusions and Future Perspectives

Over the past decade, we have gained much insight into the epigenetic modifiers that specify and maintain endocrine cell fates. Growing knowledge of the influence of diet and the maternal environment on metabolic health, the contribution of ncRNAs to islet-enriched TF function, and therapeutic applications of modification inhibitors provide foundational tools for maintaining islet cell health. With the continued development of next-generation sequencing and multiomic technologies, T1D and T2D human samples may provide a more profound knowledge of diabetes-induced chromatin and transcriptional complex changes that can be adapted for personalized medicine. These advances, however, will require further elucidation of understudied histone modifications, including histone phosphorylation and ubiquitination. Nonetheless, epigenetic modifications have proven to be crucial players in the development of the endocrine pancreas. With continued studies, they can be leveraged to revolutionize the islet biology and diabetes field.Compliance with Ethical StandardsChad Hunter declares no conflict of interest. Tanya Pierre declared no conflict of interest. Elian Toren declares no conflict of interest. Jessica Kepple declares no conflict of interest. This chapter does not contain any studies with any human participants or animals performed by any of the authors.

References

Akerman I, Tu Z, Beucher A, Rolando DMY, Sauty-Colace C, Benazra M, Nakic N, Yang J, Wang H, Pasquali L, Moran I, Garcia-Hurtado J, Castro N, Gonzalez-Franco R, Stewart AF, Bonner C, Piemonti L, Berney T, Groop L, Kerr-Conte J, Pattou F, Argmann C, Schadt E, Ravassard P, Ferrer J (2017) Human pancreatic beta cell lncRNAs control cell-specific regulatory networks. Cell Metab 25:400–411

Alejandro EU, Jo S, Akhaphong B, Llacer PR, Gianchandani M, Gregg B, Parlee SD, MacDougald OA, Bernal-Mizrachi E (2020) Maternal low-protein diet on the last week of pregnancy contributes to insulin resistance and beta-cell dysfunction in the mouse offspring. Am J Physiol Regul Integr Comp Physiol 319:R485–R496

Arnes L, Akerman I, Balderes DA, Ferrer J, Sussel L (2016) betalinc1 encodes a long noncoding RNA that regulates islet beta-cell formation and function. Genes Dev 30:502–507

Backe MB, Andersson JL, Bacos K, Christensen DP, Hansen JB, Dorosz JJ, Gajhede M, Dahlby T, Bysani M, Kristensen LH, Ling C, Olsen L, Mandrup-Poulsen T (2018) Lysine demethylase inhibition protects pancreatic beta cells from apoptosis and improves beta-cell function. Mol Cell Endocrinol 460:47–56

Backe MB, Jin C, Andreone L, Sankar A, Agger K, Helin K, Madsen AN, Poulsen SS, Bysani M, Bacos K, Ling C, Perone MJ, Holst B, Mandrup-Poulsen T (2019) The lysine demethylase KDM5B regulates islet function and glucose homeostasis. J Diabetes Res 2019:5451038

Bedford MT, Clarke SG (2009) Protein arginine methylation in mammals: who, what, and why. Mol Cell 33:1–13

Bijkerk R, Esguerra JLS, Ellenbroek JH, Au YW, Hanegraaf MAJ, de Koning EJ, Eliasson L, van Zonneveld AJ (2019) In vivo silencing of microRNA-132 reduces blood glucose and improves insulin secretion. Nucleic Acid Ther 29:67–72

Blanc RS, Richard S (2017) Arginine methylation: the coming of age. Mol Cell 65:8–24

Bompada P, Atac D, Luan C, Andersson R, Omella JD, Laakso EO, Wright J, Groop L, De Marinis Y (2016) Histone acetylation of glucose-induced thioredoxin-interacting protein gene expression in pancreatic islets. Int J Biochem Cell Biol 81:82–91

Boyson SP, Gao C, Quinn K, Boyd J, Paculova H, Frietze S, Glass KC (2021) Functional roles of bromodomain proteins in cancer. Cancers 13:3606

Campbell SA, McDonald CL, Krentz NAJ, Lynn FC, Hoffman BG (2019) TrxG complex catalytic and non-catalytic activity play distinct roles in pancreas progenitor specification and differentiation. Cell Rep 28:1830–1844 e6

Campbell SA, Begin J, McDonald CL, Vanderkruk B, Stephan TL, Hoffman BG (2021) H3K4 trimethylation is required for postnatal pancreatic endocrine cell functional maturation. Diabetes 70:2568

Carry PM, Vanderlinden LA, Johnson RK, Dong F, Steck AK, Frohnert BI, Rewers M, Yang IV, Kechris K, Norris JM (2020) DNA methylation near the INS gene is associated with INS genetic variation (rs689) and type 1 diabetes in the Diabetes Autoimmunity Study in the Young. Pediatr Diabetes 21:597–605

Cervantes S, Fontcuberta-PiSunyer M, Servitja JM, Fernandez-Ruiz R, Garcia A, Sanchez L, Lee YS, Gomis R, Gasa R (2017) Late-stage differentiation of embryonic pancreatic beta-cells requires Jarid2. Sci Rep 7:11643

Chakravarthy H, Gu X, Enge M, Dai X, Wang Y, Damond N, Downie C, Liu K, Wang J, Xing Y, Chera S, Thorel F, Quake S, Oberholzer J, MacDonald PE, Herrera PL, Kim SK (2017) Converting adult pancreatic islet alpha cells into beta cells by targeting both Dnmt1 and Arx. Cell Metab 25:622–634

Chen WB, Gao L, Wang J, Wang YG, Dong Z, Zhao J, Mi QS, Zhou L (2016) Conditional ablation of HDAC3 in islet beta cells results in glucose intolerance and enhanced susceptibility to STZ-induced diabetes. Oncotarget 7:57485–57497

Dahlby T, Simon C, Backe MB, Dahllof MS, Holson E, Wagner BK, Boni-Schnetzler M, Marzec MT, Lundh M, Mandrup-Poulsen T (2020) Enhancer of Zeste Homolog 2 (EZH2) mediates glucolipotoxicity-induced apoptosis in beta-cells. Int J Mol Sci 21:8016

Damgaard RB (2021) The ubiquitin system: from cell signalling to disease biology and new therapeutic opportunities. Cell Death Differ 28:423–426

Damond N, Thorel F, Kim SK, Herrera PL (2017) Dnmt1 activity is dispensable in delta-cells but is essential for alpha-cell homeostasis. Int J Biochem Cell Biol 88:226–235

Daneshpajooh M, Bacos K, Bysani M, Bagge A, Ottosson Laakso E, Vikman P, Eliasson L, Mulder H, Ling C (2017) HDAC7 is overexpressed in human diabetic islets and impairs insulin secretion in rat islets and clonal beta cells. Diabetologia 60:116–125

Davidson RK, Kanojia S, Spaeth JM (2021) The contribution of transcriptional coregulators in the maintenance of beta-cell function and identity. Endocrinology 162:bqaa213

denDekker AD, Davis FM, Joshi AD, Wolf SJ, Allen R, Lipinski J, Nguyen B, Kirma J, Nycz D, Bermick J, Moore BB, Gudjonsson JE, Kunkel SL, Gallagher KA (2020) TNF-alpha regulates diabetic macrophage function through the histone acetyltransferase MOF. JCI Insight 5:e132306

Dhawan S, Dirice E, Kulkarni RN, Bhushan A (2016) Inhibition of TGF-beta signaling promotes human pancreatic beta-cell replication. Diabetes 65:1208–1218

Ding H, Wang F, Shi X, Ma H, Du Y, Hou L, Xing N (2020) LncRNA MALAT1 induces the dysfunction of beta cells via reducing the histone acetylation of the PDX-1 promoter in type 1 diabetes. Exp Mol Pathol 114:104432

Draney C, Austin MC, Leifer AH, Smith CJ, Kener KB, Aitken TJ, Hess KH, Haines AC, Lett LA, Hernandez-Carretero A, Fueger PT, Arlotto M, Tessem JS (2018) HDAC1 overexpression enhances beta-cell proliferation by down-regulating Cdkn1b/p27. Biochem J 475:3997–4010

Du A, Hunter CS, Murray J, Noble D, Cai CL, Evans SM, Stein R, May CL (2009) Islet-1 is required for the maturation, proliferation, and survival of the endocrine pancreas. Diabetes 58:2059–2069

Dumasia NP, Pethe PS (2020) Pancreas development and the Polycomb group protein complexes. Mech Dev 164:103647

Ediger BN, Du A, Liu J, Hunter CS, Walp ER, Schug J, Kaestner KH, Stein R, Stoffers DA, May CL (2014) Islet-1 is essential for pancreatic beta-cell function. Diabetes 63:4206–4217

Elghazi L, Blandino-Rosano M, Alejandro E, Cras-Meneur C, Bernal-Mizrachi E (2017) Role of nutrients and mTOR signaling in the regulation of pancreatic progenitors development. Mol Metab 6:560–573

Fischle W, Wang Y, Allis CD (2003) Histone and chromatin cross-talk. Curr Opin Cell Biol 15:172–183

Fontcuberta-PiSunyer M, Cervantes S, Miquel E, Mora-Castilla S, Laurent LC, Raya A, Gomis R, Gasa R (1861) Modulation of the endocrine transcriptional program by targeting histone modifiers of the H3K27me3 mark. Biochim Biophys Acta Gene Regul Mech 2018:473–480

Golson ML (2021) Islet epigenetic impacts on beta-cell identity and function. Compr Physiol 11:1961–1978
Gong M, Yu Y, Liang L, Vuralli D, Froehler S, Kuehnen P, Du Bois P, Zhang J, Cao A, Liu Y, Hussain K, Fielitz J, Jia S, Chen W, Raile K (2019) HDAC4 mutations cause diabetes and induce beta-cell FoxO1 nuclear exclusion. Mol Genet Genomic Med 7:e602
Guarnaccia AD, Tansey WP (2018) Moonlighting with WDR5: a cellular multitasker. J Clin Med 7:21
Gujral P, Mahajan V, Lissaman AC, Ponnampalam AP (2020) Histone acetylation and the role of histone deacetylases in normal cyclic endometrium. Reprod Biol Endocrinol 18:84
Guo X, Li D, Song J, Yang Q, Wang M, Yang Y, Wang L, Hou X, Chen L, Li X (2020) Mof regulates glucose level via altering different alpha-cell subset mass and intra-islet glucagon-like peptide-1, glucagon secretion. Metabolism 109:154290
Guo X, Cui C, Song J, He Q, Zang N, Hu H, Wang X, Li D, Wang C, Hou X, Li X, Liang K, Yan F, Chen L (2021) Mof acetyltransferase inhibition ameliorates glucose intolerance and islet dysfunction of type 2 diabetes via targeting pancreatic alpha-cells. Mol Cell Endocrinol 537:111425
Hall E, Volkov P, Dayeh T, Bacos K, Ronn T, Nitert MD, Ling C (2014) Effects of palmitate on genome-wide mRNA expression and DNA methylation patterns in human pancreatic islets. BMC Med 12:103
Hall E, Dekker Nitert M, Volkov P, Malmgren S, Mulder H, Bacos K, Ling C (2018) The effects of high glucose exposure on global gene expression and DNA methylation in human pancreatic islets. Mol Cell Endocrinol 472:57–67
He ZX, Wei BF, Zhang X, Gong YP, Ma LY, Zhao W (2021) Current development of CBP/p300 inhibitors in the last decade. Eur J Med Chem 209:112861
Hojfeldt JW, Hedehus L, Laugesen A, Tatar T, Wiehle L, Helin K (2019) Non-core subunits of the PRC2 complex are collectively required for its target-site specificity. Mol Cell 76:423–436 e3
Horwitz E, Krogvold L, Zhitomirsky S, Swisa A, Fischman M, Lax T, Dahan T, Hurvitz N, Weinberg-Corem N, Klochendler A, Powers AC, Brissova M, Jorns A, Lenzen S, Glaser B, Dahl-Jorgensen K, Dor Y (2018) Beta-cell DNA damage response promotes islet inflammation in type 1 diabetes. Diabetes 67:2305–2318
Huijbregts L, Petersen MBK, Berthault C, Hansson M, Aiello V, Rachdi L, Grapin-Botton A, Honore C, Scharfmann R (2019) Bromodomain and extra terminal protein inhibitors promote pancreatic endocrine cell fate. Diabetes 68:761–773
Hyun K, Jeon J, Park K, Kim J (2017) Writing, erasing and reading histone lysine methylations. Exp Mol Med 49:e324
Inoue H, Asahara SI, Sugiura Y, Kawada Y, Imai A, Hara C, Kanno A, Kimura-Koyanagi M, Kido Y (2021) Histone deacetylase 6 regulates insulin signaling in pancreatic beta cells. Biochem Biophys Res Commun 534:896–901
Jeon JP, Koh IU, Choi NH, Kim BJ, Han BG, Lee S (2017) Differential DNA methylation of MSI2 and its correlation with diabetic traits. PLoS One 12:e0177406
Johnson RK, Vanderlinden LA, Dong F, Carry PM, Seifert J, Waugh K, Shorrosh H, Fingerlin T, Frohnert BI, Yang IV, Kechris K, Rewers M, Norris JM (2020) Longitudinal DNA methylation differences precede type 1 diabetes. Sci Rep 10:3721
Jonsson J, Carlsson L, Edlund T, Edlund H (1994) Insulin-promoter-factor 1 is required for pancreas development in mice. Nature 371:606–609
Jorgensen MC, Ahnfelt-Ronne J, Hald J, Madsen OD, Serup P, Hecksher-Sorensen J (2007) An illustrated review of early pancreas development in the mouse. Endocr Rev 28:685–705
Kameswaran V, Bramswig NC, McKenna LB, Penn M, Schug J, Hand NJ, Chen Y, Choi I, Vourekas A, Won KJ, Liu C, Vivek K, Naji A, Friedman JR, Kaestner KH (2014) Epigenetic regulation of the DLK1-MEG3 microRNA cluster in human type 2 diabetic islets. Cell Metab 19:135–145
Kartha RV, Subramanian S (2014) Competing endogenous RNAs (ceRNAs): new entrants to the intricacies of gene regulation. Front Genet 5:8

Kawada Y, Asahara SI, Sugiura Y, Sato A, Furubayashi A, Kawamura M, Bartolome A, Terashi-Suzuki E, Takai T, Kanno A, Koyanagi-Kimura M, Matsuda T, Hashimoto N, Kido Y (2017) Histone deacetylase regulates insulin signaling via two pathways in pancreatic beta cells. PLoS One 12:e0184435

Kim JK, Lim Y, Lee JO, Lee YS, Won NH, Kim H, Kim HS (2015) PRMT4 is involved in insulin secretion via the methylation of histone H3 in pancreatic beta cells. J Mol Endocrinol 54:315–324

Kim H, Yoon BH, Oh CM, Lee J, Lee K, Song H, Kim E, Yi K, Kim MY, Kim H, Kim YK, Seo EH, Heo H, Kim HJ, Lee J, Suh JM, Koo SH, Seong JK, Kim S, Ju YS, Shong M, Kim M, Kim H (2020) PRMT1 is required for the maintenance of mature beta-cell identity. Diabetes 69:355–368

Kong X, Liu CX, Wang GD, Yang H, Yao XM, Hua Q, Li XY, Zhang HM, Ma MZ, Su Q, Lv K (2019) LncRNA LEGLTBC functions as a ceRNA to antagonize the effects of miR-34a on the downregulation of SIRT1 in glucolipotoxicity-induced INS-1 beta cell oxidative stress and apoptosis. Oxidative Med Cell Longev 2019:4010764

Kuo LJ, Yang LX (2008) Gamma-H2AX - a novel biomarker for DNA double-strand breaks. In Vivo 22:305–309

Lee JE, Park YK, Park S, Jang Y, Waring N, Dey A, Ozato K, Lai B, Peng W, Ge K (2017) Brd4 binds to active enhancers to control cell identity gene induction in adipogenesis and myogenesis. Nat Commun 8:2217

Lee K, Kim H, Lee J, Oh CM, Song H, Kim H, Koo SH, Lee J, Lim A, Kim H (2019) Essential role of protein arginine methyltransferase 1 in pancreas development by regulating protein stability of neurogenin 3. Diabetes Metab J 43:649–658

Li J, Liu C (2019) Coding or noncoding, the converging concepts of RNAs. Front Genet 10:496

Li G, Petkova TD, Laritsky E, Kessler N, Baker MS, Zhu S, Waterland RA (2019) Early postnatal overnutrition accelerates aging-associated epigenetic drift in pancreatic islets. Environ Epigenet 5:dvz015

Li N, Jiang D, He Q, He F, Li Y, Deng C, Li F (2020) microRNA-181c-5p promotes the formation of insulin-producing cells from human induced pluripotent stem cells by targeting smad7 and TGIF2. Cell Death Dis 11:462

Liu Y, Tan Q, Liu F (2018) Differentially methylated circulating DNA: a novel biomarker to monitor beta cell death. J Diabetes Complicat 32:349–353

Liu Z, Ding Z, Lynch EC, McCauley N, Zhou Y, Zhang KK, Xie L (2020) Pregestational diet transition to normal-fat diet avoids the deterioration of pancreatic beta-cell function in male offspring induced by maternal high-fat diet. J Nutr Biochem 86:108495

Lu TT, Heyne S, Dror E, Casas E, Leonhardt L, Boenke T, Yang CH, Sagar L, Arrigoni K, Dalgaard R, Teperino L, Enders M, Selvaraj M, Ruf SJ, Raja H, Xie U, Boenisch SH, Orkin FC, Lynn BG, Hoffman D, Grun T, Vavouri AML, Pospisilik JA (2018) The polycomb-dependent epigenome controls beta cell dysfunction, dedifferentiation, and diabetes. Cell Metab 27:1294–1308 e7

Luo C, Hajkova P, Ecker JR (2018) Dynamic DNA methylation: in the right place at the right time. Science 361:1336–1340

Ma J, He X, Cao Y, O'Dwyer K, Szigety KM, Wu Y, Gurung B, Feng Z, Katona BW, Hua X (2020) Islet-specific Prmt5 excision leads to reduced insulin expression and glucose intolerance in mice. J Endocrinol 244:41–52

Makkar R, Behl T, Arora S (2020) Role of HDAC inhibitors in diabetes mellitus. Curr Res Transl Med 68:45–50

Martinez-Sanchez A, Rutter GA, Latreille M (2016) MiRNAs in beta-cell development, identity, and disease. Front Genet 7:226

McCann J, Ellis M, McGee SL, Aston-Mourney K (2019) Class IIa HDACs do not influence beta-cell function under normal or high glucose conditions. Islets 11:112–118

Millan-Zambrano G, Santos-Rosa H, Puddu F, Robson SC, Jackson SP, Kouzarides T (2018) Phosphorylation of histone H4T80 triggers DNA damage checkpoint recovery. Mol Cell 72:625–635 e4

Muhammad AB, Xing B, Liu C, Naji A, Ma X, Simmons RA, Hua X (2017) Menin and PRMT5 suppress GLP1 receptor transcript and PKA-mediated phosphorylation of FOXO1 and CREB. Am J Physiol Endocrinol Metab 313:E148–E166

Mziaut H, Henniger G, Ganss K, Hempel S, Wolk S, McChord J, Chowdhury K, Ravassard P, Knoch KP, Krautz C, Weitz J, Grutzmann R, Pilarsky C, Solimena M, Kersting S (2020) MiR-132 controls pancreatic beta cell proliferation and survival through Pten/Akt/Foxo3 signaling. Mol Metab 31:150–162

Nakagawa T, Xiong Y (2011) X-linked mental retardation gene CUL4B targets ubiquitylation of H3K4 methyltransferase component WDR5 and regulates neuronal gene expression. Mol Cell 43:381–391

O'Brien J, Hayder H, Zayed Y, Peng C (2018) Overview of MicroRNA biogenesis, mechanisms of actions, and circulation. Front Endocrinol 9:402

Park SY, Kim JS (2020) A short guide to histone deacetylases including recent progress on class II enzymes. Exp Mol Med 52:204–212

Parveen N, Dhawan S (2021) DNA methylation patterning and the regulation of beta cell homeostasis. Front Endocrinol 12:651258

Qian B, Yang Y, Tang N, Wang J, Sun P, Yang N, Chen F, Wu T, Sun T, Li Y, Chang X, Zhu Y, Zhang Y, Han X (2021) M1 macrophage-derived exosomes impair beta cell insulin secretion via miR-212-5p by targeting SIRT2 and inhibiting Akt/GSK-3beta/beta-catenin pathway in mice. Diabetologia 64:2037–2051

Qin K, Zhang N, Zhang Z, Nipper M, Zhu Z, Leighton J, Xu K, Musi N, Wang P (2018) SIRT6-mediated transcriptional suppression of Txnip is critical for pancreatic beta cell function and survival in mice. Diabetologia 61:906–918

Rachdi L, Aiello V, Duvillie B, Scharfmann R (2012) L-leucine alters pancreatic beta-cell differentiation and function via the mTor signaling pathway. Diabetes 61:409–417

Remsberg JR, Ediger BN, Ho WY, Damle M, Li Z, Teng C, Lanzillotta C, Stoffers DA, Lazar MA (2017) Deletion of histone deacetylase 3 in adult beta cells improves glucose tolerance via increased insulin secretion. Mol Metab 6:30–37

Rewers M, Ludvigsson J (2016) Environmental risk factors for type 1 diabetes. Lancet 387:2340–2348

Rossetto D, Avvakumov N, Cote J (2012) Histone phosphorylation: a chromatin modification involved in diverse nuclear events. Epigenetics 7:1098–1108

Ruiz L, Gurlo T, Ravier MA, Wojtusciszyn A, Mathieu J, Brown MR, Broca C, Bertrand G, Butler PC, Matveyenko AV, Dalle S, Costes S (2018) Proteasomal degradation of the histone acetyl transferase p300 contributes to beta-cell injury in a diabetes environment. Cell Death Dis 9:600

Schuettengruber B, Bourbon HM, Di Croce L, Cavalli G (2017) Genome regulation by polycomb and trithorax: 70 years and counting. Cell 171:34–57

Sebastiani G, Valentini M, Grieco GE, Ventriglia G, Nigi L, Mancarella F, Pellegrini S, Martino G, Sordi V, Piemonti L, Dotta F (2017) MicroRNA expression profiles of human iPSCs differentiation into insulin-producing cells. Acta Diabetol 54:265–281

Shimizu I, Yoshida Y, Suda M, Minamino T (2014) DNA damage response and metabolic disease. Cell Metab 20:967–977

Singer RA, Arnes L, Cui Y, Wang J, Gao Y, Guney MA, Burnum-Johnson KE, Rabadan R, Ansong C, Orr G, Sussel L (2019) The long noncoding RNA paupar modulates PAX6 regulatory activities to promote alpha cell development and function. Cell Metab 30:1091–1106 e8

Smeenk G, Mailand N (2016) Writers, readers, and erasers of histone ubiquitylation in DNA double-strand break repair. Front Genet 7:122

Solorzano-Vargas RS, Bjerknes M, Wu SV, Wang J, Stelzner M, Dunn JCY, Dhawan S, Cheng H, Georgia S, Martin MG (2019) The cellular regulators PTEN and BMI1 help mediate NEUROGENIN-3-induced cell cycle arrest. J Biol Chem 294:15182–15192

Song MY, Wang J, Ka SO, Bae EJ, Park BH (2016) Insulin secretion impairment in Sirt6 knockout pancreatic beta cells is mediated by suppression of the FoxO1-Pdx1-Glut2 pathway. Sci Rep 6:30321

Sun Q, Xu H, Xue J, Yang Q, Chen C, Yang P, Han A, Tu Q, Lu J, Gao X, Xiang Q, Liu Q (2018) MALAT1 via microRNA-17 regulation of insulin transcription is involved in the dysfunction of pancreatic beta-cells induced by cigarette smoke extract. J Cell Physiol 233:8862–8873
Tay VSY, Devaraj S, Koh T, Ke G, Crasta KC, Ali Y (2019) Increased double strand breaks in diabetic beta-cells with a p21 response that limits apoptosis. Sci Rep 9:19341
Tinajero MG, Malik VS (2021) An update on the epidemiology of type 2 diabetes: a global perspective. Endocrinol Metab Clin N Am 50:337–355
Uckelmann M, Sixma TK (2017) Histone ubiquitination in the DNA damage response. DNA Repair 56:92–101
Van de Velde S, Wiater E, Tran M, Hwang Y, Cole PA, Montminy M (2019) CREB promotes Beta cell gene expression by targeting its coactivators to tissue-specific enhancers. Mol Cell Biol 39:e00200
Vinckier NK, Patel NA, Geusz RJ, Wang A, Wang J, Matta I, Harrington AR, Wortham M, Wetton N, Wang J, Jhala US, Rosenfeld MG, Benner CW, Shih HP, Sander M (2020) LSD1-mediated enhancer silencing attenuates retinoic acid signalling during pancreatic endocrine cell development. Nat Commun 11:2082
Volkmar M, Dedeurwaerder S, Cunha DA, Ndlovu MN, Defrance M, Deplus R, Calonne E, Volkmar U, Igoillo-Esteve M, Naamane N, Del Guerra S, Masini M, Bugliani M, Marchetti P, Cnop M, Eizirik DL, Fuks F (2012) DNA methylation profiling identifies epigenetic dysregulation in pancreatic islets from type 2 diabetic patients. EMBO J 31:1405–1426
Volkov P, Bacos K, Ofori JK, Esguerra JL, Eliasson L, Ronn T, Ling C (2017) Whole-genome bisulfite sequencing of human pancreatic islets reveals novel differentially methylated regions in type 2 diabetes pathogenesis. Diabetes 66:1074–1085
Wade AK, Liu Y, Bethea MM, Toren E, Tse HM, Hunter CS (2019) LIM-domain transcription complexes interact with ring-finger ubiquitin ligases and thereby impact islet beta-cell function. J Biol Chem 294:11728–11740
Wang N, Zhu Y, Xie M, Wang L, Jin F, Li Y, Yuan Q, De W (2018) Long noncoding RNA Meg3 regulates mafa expression in mouse beta cells by inactivating Rad21, Smc3 or Sin3alpha. Cell Physiol Biochem 45:2031–2043
Wang Y, Sun J, Ni Q, Nie A, Gu Y, Wang S, Zhang W, Ning G, Wang W, Wang Q (2019) Dual effect of raptor on neonatal beta-cell proliferation and identity maintenance. Diabetes 68:1950–1964
Wei JW, Huang K, Yang C, Kang CS (2017) Non-coding RNAs as regulators in epigenetics (review). Oncol Rep 37:3–9
Whyte WA, Bilodeau S, Orlando DA, Hoke HA, Frampton GM, Foster CT, Cowley SM, Young RA (2012) Enhancer decommissioning by LSD1 during embryonic stem cell differentiation. Nature 482:221–225
Wong CK, Wade-Vallance AK, Luciani DS, Brindle PK, Lynn FC, Gibson WT (2018) The p300 and CBP transcriptional coactivators are required for beta-cell and alpha-cell proliferation. Diabetes 67:412–422
Wortham M, Sander M (2021) Transcriptional mechanisms of pancreatic beta-cell maturation and functional adaptation. Trends Endocrinol Metab 32:474–487
Xie Z, Chang C, Huang G, Zhou Z (2020) The role of epigenetics in type 1 diabetes. Adv Exp Med Biol 1253:223–257
Xiong X, Wang G, Tao R, Wu P, Kono T, Li K, Ding WX, Tong X, Tersey SA, Harris RA, Mirmira RG, Evans-Molina C, Dong XC (2016a) Sirtuin 6 regulates glucose-stimulated insulin secretion in mouse pancreatic beta cells. Diabetologia 59:151–160
Xiong X, Sun X, Wang Q, Qian X, Zhang Y, Pan X, Dong XC (2016b) SIRT6 protects against palmitate-induced pancreatic beta-cell dysfunction and apoptosis. J Endocrinol 231:159–165
Yajnik CS, Deshpande SS, Jackson AA, Refsum H, Rao S, Fisher DJ, Bhat DS, Naik SS, Coyaji KJ, Joglekar CV, Joshi N, Lubree HG, Deshpande VU, Rege SS, Fall CH (2008) Vitamin B12 and folate concentrations during pregnancy and insulin resistance in the offspring: the Pune Maternal Nutrition Study. Diabetologia 51:29–38

Yang XF, Zhou SY, Wang C, Huang W, Li N, He F, Li FR (2020) Inhibition of LSD1 promotes the differentiation of human induced pluripotent stem cells into insulin-producing cells. Stem Cell Res Ther 11:185
Yu XX, Qiu WL, Yang L, Li LC, Zhang YW, Xu CR (2018) Dynamics of chromatin marks and the role of JMJD3 during pancreatic endocrine cell fate commitment. Development 145:dev163162
Zhang C, Moriguchi T, Kajihara M, Esaki R, Harada A, Shimohata H, Oishi H, Hamada M, Morito N, Hasegawa K, Kudo T, Engel JD, Yamamoto M, Takahashi S (2005) MafA is a key regulator of glucose-stimulated insulin secretion. Mol Cell Biol 25:4969–4976
Zhang T, Cooper S, Brockdorff N (2015) The interplay of histone modifications - writers that read. EMBO Rep 16:1467–1481
Zhang HH, Ma XJ, Wu LN, Zhao YY, Zhang PY, Zhang YH, Shao MW, Liu F, Li F, Qin GJ (2016) Sirtuin-3 (SIRT3) protects pancreatic beta-cells from endoplasmic reticulum (ER) stress-induced apoptosis and dysfunction. Mol Cell Biochem 420:95–106
Zhang Y, Thai K, Jin T, Woo M, Gilbert RE (2018) SIRT1 activation attenuates alpha cell hyperplasia, hyperglucagonaemia and hyperglycaemia in STZ-diabetic mice. Sci Rep 8:13972
Zhang Y, Zhou F, Bai M, Liu Y, Zhang L, Zhu Q, Bi Y, Ning G, Zhou L, Wang X (2019) The pivotal role of protein acetylation in linking glucose and fatty acid metabolism to beta-cell function. Cell Death Dis 10:66
Zhang Y, Wang S, Zhang L, Zhou F, Zhu K, Zhu Q, Liu Q, Liu Y, Jiang L, Ning G, Bi Y, Zhou L, Wang X (2020a) Protein acetylation derepresses serotonin synthesis to potentiate pancreatic Beta-cell function through HDAC1-PKA-Tph1 signaling. Theranostics 10:7351–7368
Zhang FF, Liu YH, Wang DW, Liu TS, Yang Y, Guo JM, Pan Y, Zhang YF, Du H, Li L, Jin L (2020b) Obesity-induced reduced expression of the lncRNA ROIT impairs insulin transcription by downregulation of Nkx6.1 methylation. Diabetologia 63:811–824
Zhang L, Sheng C, Zhou F, Zhu K, Wang S, Liu Q, Yuan M, Xu Z, Liu Y, Lu J, Liu J, Zhou L, Wang X (2021a) CBP/p300 HAT maintains the gene network critical for beta cell identity and functional maturity. Cell Death Dis 12:476
Zhang F, Yang Y, Chen X, Liu Y, Hu Q, Huang B, Liu Y, Pan Y, Zhang Y, Liu D, Liang R, Li G, Wei Q, Li L, Jin L (2021b) The long non-coding RNA betaFaar regulates islet beta-cell function and survival during obesity in mice. Nat Commun 12:3997
Zhong A, Chang M, Yu T, Gau R, Riley DJ, Chen Y, Chen PL (2018) Aberrant DNA damage response and DNA repair pathway in high glucose conditions. J Can Res Updates 7:64–74
Zhou Q, Law AC, Rajagopal J, Anderson WJ, Gray PA, Melton DA (2007) A multipotent progenitor domain guides pancreatic organogenesis. Dev Cell 13:103–114
Zhou S, Tang X, Chen HZ (2018) Sirtuins and insulin resistance. Front Endocrinol 9:748
Zhou AX, Mondal T, Tabish AM, Abadpour S, Ericson E, Smith DM, Knoll R, Scholz H, Kanduri C, Tyrberg B, Althage M (2021) The long noncoding RNA TUNAR modulates Wnt signaling and regulates human beta-cell proliferation. Am J Physiol Endocrinol Metab 320:E846–E857

Development of the Pancreatic Ducts and Their Contribution to Organogenesis

Dorian J. Dale, Caleb D. Rutan, and Teresa L. Mastracci

Abstract The pancreas is a dual-function organ, with exocrine cells that aid in digestion and endocrine cells that regulate glucose homeostasis. These cell types share common progenitors and arise from the embryonic ducts. Early signaling events in the embryonic ducts shape the neonatal, adolescent, and adult exocrine and endocrine pancreas. This chapter discusses recent advances in the tools used to study the ducts and our current understanding of how ductal development contributes to pancreatic organogenesis.

Keywords Acinar · Duct · Pancreas · Progenitor

Abbreviations

AMY	Amylase
CPA1	Carboxypeptidase 1
CDC42	Cell division cycle 42
CF	Cystic fibrosis

D. J. Dale · C. D. Rutan
Department of Biology, Indiana University-Indianapolis,
Indianapolis, IN, USA
e-mail: dorjdale@iu.edu; cdrutan@iu.edu

T. L. Mastracci (✉)
Department of Biology, Indiana University-Purdue University-Indianapolis,
Indianapolis, IN, USA

Department of Biochemistry and Molecular Biology, Indiana University School of Medicine,
Indianapolis, IN, USA

Center for Diabetes and Metabolic Disease, Indiana University School of Medicine,
Indianapolis, IN, USA
e-mail: tmastrac@iu.edu

M. L. Golson (ed.), *Molecular and Cell Biology of Pancreas Development, Function and Regeneration*, Advances in Anatomy, Embryology and Cell Biology 239, https://doi.org/10.1007/978-3-031-62232-8_2

CFTR	CF transmembrane conductance receptor
ECM	Extracellular matrix
FGF	Fibroblast growth factor
FACS	Fluorescence-activated cell sorting
GLIS3	Glis family Zinc Finger 3
HNF	Hepatocyte nuclear factor
HES1	Hairy/Enhancer of Split 1
hESCs	Human embryonic stem cells
IHH	Indian hedgehog
MACD	Mitosis-associated cell dispersal
MUC1	Mucin1
MPCs	Multipotent progenitor cells
c-MYC	Myc proto-oncogene
NGN3	Neurogenin 3
NKX 6.1	NK homeobox 6.1
NICD	Notch intracellular domain
PDX1	Pancreas duodenal homeobox 1
PTF1A	Pancreas transcription factor 1a
PAR6	Partitioning defective 6
pMLC	Phospho-myosin light chain complex
PCP	Planar cell polarity
PKD	Polycystic Kidney Disease
RBP-J	Recombination signaling protein for Immunoglobulin Kappa J
SHH	Sonic hedgehog
SOX	SRY-box
TGF-β	Transforming growth factor β
T3cD	Type 3c diabetes
Vangl	VANGL planar cell polarity proteins
YAP1	Yes1-associated transcription factor

1 Introduction

The exocrine pancreas is composed of acinar cells and duct cells, which together coordinate a critical digestion function (Domínguez-Muñoz 2011; Bastidas-Ponce et al. 2017; Larsen and Grapin-Botton 2017). This functional relationship between the acinar cells and the ducts is possible due to the physical location of the cell types relative to each other. Specifically, the acinar cells produce zymogens (proenzymes) that are packaged into granules and secreted into the ducts for their subsequent transport to the intestine to aid in digestion (Domínguez-Muñoz 2011). The ductal cells synthesize and secrete bicarbonate, which maintains these digestive proenzymes in an inactive state. Disruptions to exocrine function due to altered development, organization, or function of the ducts can result in disease (Ehre et al. 2014; Hegyi and Sahin-Tóth 2017; Soave et al. 2014). Concordantly, the ductal system also plays a critical role in the development and function of the endocrine portion of the pancreas. The islets of Langerhans are composed of hormone-producing endocrine cells (α, β, δ, ε, and PP cells). The progenitor cells from which all islet cell types are derived originate in the embryonic ducts. Therefore, the ducts play a critical role in the development, organization, and function of the whole pancreas. In this chapter, we will review our current understanding as well as recent advances in (a) pancreatic progenitor specification and duct cell identity; (b) structure and formation of the pancreatic epithelium; (c) tools for understanding duct organization/development; and (d) the role of the pancreatic ducts in organ function and dysfunction.

1.1 Specification of the Pancreatic Progenitors and Duct Cells

1.1.1 Pancreatic Progenitor Cell Specification

The pancreas originates from the dorsal and ventral foregut endoderm as two bud-like protrusions, which are specified beginning at embryonic day (E) 9 (Deutsch et al. 2001). The dorsal bud evaginates first, followed by ventral bud evagination in the subsequent 12–24 h (Haumaitre et al. 2005; Bort et al. 2004). The epithelial cells that compose this early domain (the pancreatic progenitor cells) are marked by the expression of specific transcription factors, namely, pancreas duodenal homeobox 1 (*Pdx1*) and pancreas transcription factor 1a (*Ptf1a*) (Haumaitre et al. 2005). *Pdx1* and *Ptf1a* co-expression is necessary and sufficient for pancreas development (Kawaguchi et al. 2002; Burlison et al. 2008). Together, these transcription factors are responsible for initiating the genetic programs required for pancreatic organogenesis. The generation of mutant mouse models has illustrated the importance of these and many other transcription factors (Table 1 summarizes select mouse models related to pancreas and duct development).

First and foremost, deletion of *Pdx1* was reported to result in complete pancreatic agenesis, which solidified its critical role in pancreas specification (Jonsson et al. 1994). *Ptf1a*$^{-/-}$ mice also showed pancreatic agenesis (Kawaguchi et al. 2002; Krapp et al. 1998). Moreover, when both factors were deleted in a double-knockout model, the resulting phenotype was indistinguishable from the *Pdx1*$^{-/-}$mutant (Burlison et al. 2008). These observations were significant for identifying and clarifying the role of *Pdx1* and *Ptf1a* in cell fate decisions in the developing pancreas. Specifically, the *Pdx1*$^{-/-}$mutants showed a complete absence of all pancreatic cells, including the endocrine cells.

In contrast, the *Ptf1a*$^{-/-}$ mutants also showed pancreatic agenesis but maintained a small population of endocrine cells near the dorsal pancreatic ruminant (Burlison et al. 2008). The genetic program initiated by co-expression of *Pdx1* and *Ptf1a* within the pancreatic progenitors ultimately instructs the expression of numerous transcription factors, which maintain the pancreatic progenitor cell population and drive the progenitors to differentiate into all pancreatic cell types (Zhou et al. 2007).

Table 1 Mouse models of pancreatic ductal development

Gene	Genotype of Mouse Allele(s)	Phenotype	Ref.
Pdx1	$Pdx1^{-/-}$ $Pdx1^{LacZ/LacZ}$	Pancreatic agenesis; microlumen impairment; reduction in proliferation; cell polarity defects	(Jonsson et al. 1994; Marty-Santos and Cleaver 2016)
Ptf1a	$Ptf1a^{-/-}$ $Ptf1a^{Cre/Cre}$;*R26R-LacZ* $Ptf1a^{YFP/YFP}$	Pancreatic agenesis with residual endocrine cells	(Kawaguchi et al. 2002; Burlison et al. 2008; Krapp et al. 1998)
Pdx1;Ptf1a	$Ptf1a^{YFP/YFP}$;$Pdx1^{LacZ/LacZ}$	Pancreatic agenesis	(Burlison et al. 2008)
Hnf1β	*Pdx1-Cre*;$Hnf1\beta^{loxP/loxP}$ *Sox9-*Cre^{ER};$Hnf1\beta^{loxP/loxP}$	Ductal cysts. Loss of cilia	(De Vas et al. 2015)
Gata4/ Gata6	*Pdx1-Cre*;$Gata4^{loxP/loxP}$;$Gata6^{loxP/loxP}$	Arrested pancreas development; liver and stomach cell fates specification	(Xuan and Sussel 2016)
Gata4	$Gata4^{-/-}$	Ventral pancreatic agenesis	(Watt et al. 2007)
Glis3	$Glis3^{mCherry/mCherry}$	Ductal cysts	(Kang et al. 2016)
Hnf6	*Pdx1-Cre*;$Hnf6^{loxP/loxP}$	Ductal cysts	(Zhang et al. 2009)
Rbp-j	*Pdx1-Cre*; *Rbp-*$j^{loxP/loxP}$	Decreased branching; decreased pancreas size and β-cell mass	(Fujikura et al. 2006)
Maml1	*Hnf1β-*Cre^{ERT2};$Rosa26^{dn\,Maml1}$	Cell fate conversion of bipotent trunk cells to tip cells; precocious endocrine differentiation concomitant with loss of duct cells	(Seymour et al. 2020)
Alfadin	*Pdx1-Cre*;$Alfadin^{loxP/loxP}$ *Pdx1-*Cre^{ER};$Alfadin^{loxP/loxP}$ *Sox9-*Cre^{ER};$Alfadin^{loxP/loxP}$	Lumen formation failure; uncoordinated apical membrane localization	(Azizoglu et al. 2017)
Itgb1	*Pdx1-Cre*;$Itgb1^{loxP/loxP}$	Impaired branching	(Shih et al. 2016)
Cdc42	*Pdx1-Cre*;$Cdc4^{loxP/loxP}$	Loss of apical polarity and failure to maintain lumen structure. Impaired differentiation	(Kesavan et al. 2009)
Yap	*Pdx1-Cre*;$Yap^{loxP/loxP}$	Hypoglycemia; increased serum insulin; increased insulin and glucagon mRNA Increased beta cell mass	(Mamidi et al. 2018)
RhoA	*Pdx1-*Cre^{ER};$RhoA^{loxP/loxP}$	Impaired lumen network formation; uncoordinated apical membrane localization	(Azizoglu et al. 2017)
Vangl2	*Pdx1-Cre*;*Tg(pCAGGS-LoxP-cherry-Vgl2)*^{AGB}	Cell death and pancreatic hypoplasia	(Flasse et al. 2020)
Rosa26	*Rosa26-*Cre^{ERT2}/$R26R^{Confetti}$	Inducible reporter allele used to label clonal populations of duct cells	(Sznurkowska et al. 2018)

Table 1 (continued)

Gene	Genotype of Mouse Allele(s)	Phenotype	Ref.
Cftr	$MIPCre^{ERT};Cftr^{loxP}/Cftr^{\Delta 11}$	Acinar apoptosis; fibrosis; lipomatosis; cystic ducts	(Hart et al. 2018)
Ift88	$Ift88^{delta2\text{-}3betagal}$ $Tg737^{orpk}$	Acinar vacuolization; acinar ablation; hypoglycemia	(Zhang et al. 2005)
Lkb1	$Sox9\text{-}Cre^{ER};Lkb1^{loxP/loxP}$	Enlarged ducts; acinar apoptosis; acinar-ductal metaplasia; and lipomatosis	(Augereau et al. 2016)
Hnf6	$Sox9\text{-}Cre^{ER};Hnf6^{loxP/-}$	Enlarged ducts; acinar apoptosis; acinar-ductal metaplasia; lipomatosis	(Augereau et al. 2016)

For example, the co-expression of SRY-box 9 (SOX9) in the PDX1-expressing foregut progenitor cells is critical for driving these cells toward a pancreas cell fate (Shih et al. 2015). Similarly, the expression of hepatocyte nuclear factor 1β (HNF1β) in the multipotent pancreatic progenitors specifies a subset of cells in the ductal epithelium that become bipotent, leading to either an endocrine or ductal cell fate (Haumaitre et al. 2005; Zhang et al. 2009; Solar et al. 2009). These are just two examples that demonstrate how the ductal epithelium is essential for the development of all pancreatic lineages.

Signaling from surrounding tissues also assists with the specification of the pancreatic progenitor cells. One of these early critical signaling relationships is between fibroblast growth factor (FGF) and sonic hedgehog (SHH). Hebrok et al. (Hebrok et al. 1998) demonstrated that FGF2 signaling from the notochord repressed expression of SHH in the dorsal pancreatic domain; repression of the SHH signal drove the induction of the pancreatic progenitor cells (Hebrok et al. 1998). Additional studies implicated transforming growth factor β (TGF-β) signaling in this early SHH repression (Hebrok et al. 2000), as well as the expression of the GATA transcription factors (so-named after their propensity to bind GATA DNA sequences) to maintain commitment to the pancreatic cell fate (Xuan and Sussel 2016). Interestingly, genetic deletion of both *Gata4* and *Gata6* resulted in upregulation of SHH and Indian hedgehog (IHH) in the early pancreatic ductal epithelium and downregulation of transcription factors (*Pdx1*, Neurogenin 3 [*Ngn3*], NK homeobox 6.1 [*Nkx6.1*]) associated with specification of the pancreatic cell fate (Xuan and Sussel 2016). Concomitantly, upregulation of factors specifying the intestinal and stomach cell lineages occurs. Although no clear changes in organ size were observed, these results suggest a cell fate change in the early pancreas epithelium illustrated by expression of stomach (*Sox2*) and intestinal (caudal type homeobox 2 [*Cdx2*]) genes. Clearly, early signals from the notochord directly influence the specification of the pancreatic progenitor cells and, therefore, proper patterning of the ductal epithelium.

Ventral pancreas progenitor specification also requires signals from a surrounding tissue, namely the cardiac mesoderm. The ventral pancreas derives from a bipotent pool of cells capable of differentiating toward a liver or pancreas fate. Interestingly, FGF signaling plays a role in this cell fate decision. FGF secreted from the cardiac mesoderm induces SHH signaling, inhibiting pancreatic cell fate and promoting liver specification (Deutsch et al. 2001). The proximity of the ventral pancreatic domain to the septum transversum mesenchyme is also vital for progenitor cell specification. This understanding was gained in studies analyzing *Gata4*$^{-/-}$ embryos, wherein the absence of the septum transversum mesenchyme and a failure of pancreatic progenitor cell induction were observed, resulting in agenesis of the ventral pancreas. In contrast, dorsal pancreas development proceeded normally (Watt et al. 2007). Once again, these studies provide evidence that early signaling cascades from surrounding tissues are necessary for the specification and maintenance of the pancreas progenitor cell pool, which is critical for ductal epithelium morphogenesis and all subsequent pancreatic lineages.

Overall, establishing the pancreas progenitor cells depends on intrinsic transcription factor expression and extrinsic signals from the surrounding tissues. Together, these events initiate the programs that, among many roles, instruct the specification of the ductal cell fate and morphogenesis of the epithelium, both of which are critical events in pancreas development.

1.1.2 Duct Cell Identity

The pancreatic multipotent progenitor cells (MPCs) are specified during the first stage of pancreas development, known as the "primary transition," which is also the stage when the pancreatic epithelium is established and organized (Zhou et al. 2007). Specifically, the MPCs become restricted to either a "tip" or "trunk" cell fate; these designations come from the spatial organization of the duct cells within the epithelium (Zhou et al. 2007). As visualized in Fig. 1, the *trunk* epithelium contains bipotent progenitor cells that will differentiate into either endocrine or duct cells. In contrast, the cells at the *tips* of the ducts are unipotent cells that will differentiate into exocrine acinar cells (Fig. 1a–e).

The tip and trunk ductal cells vary not only in their spatial location but also in their transcription factor expression patterns. For example, trunk cells highly express NKX6.1, while tip cells express PTF1a (Burlison et al. 2008). PTF1a expression is progressively restricted to the acinar cells in the exocrine compartment and disallowed in the endocrine and ductal cell progenitors (Pan et al. 2013). These two transcription factors act antagonistically so that the presence of NKX6.1 represses PTF1a and thereby inhibits the acinar specification (tip) program in the trunk cells (Schaffer et al. 2010). Concurrently, PTF1a acts reciprocally to inhibit the endocrine/duct specification (trunk) program in the tip cells. These observations assist our understanding of how the tip and trunk cells give rise to different mature pancreatic cell types.

Clearly, the patterning of tip and trunk cells during the primary transition involves the upregulation of not just one but many genes (Fig. 1f). Specifically, the trunk cells, which can differentiate toward either a duct or endocrine fate, can express

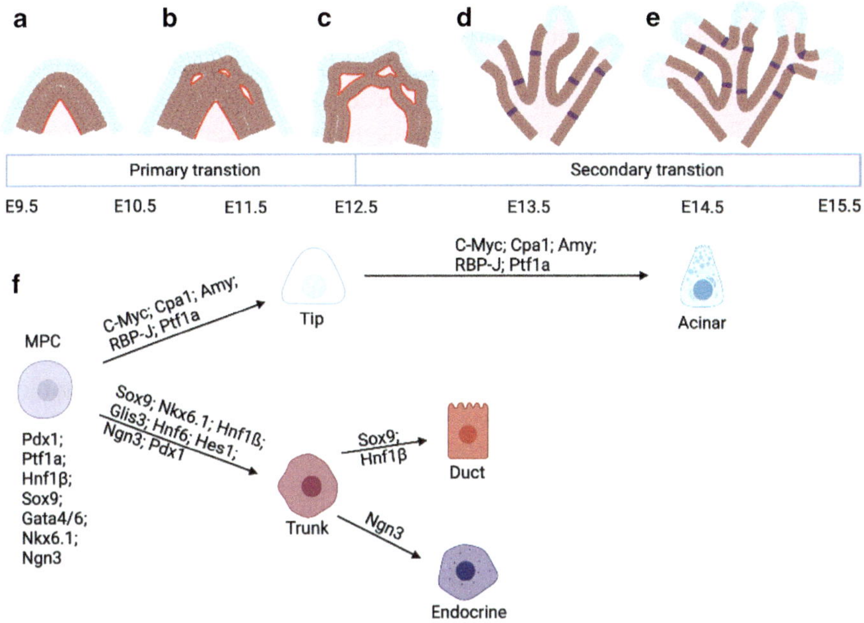

Fig. 1 Overview of pancreas development. (**a**) Pancreatic bud comprised of multipotent progenitor cells (MPCs) expanding from the foregut epithelium. Blue cells are cap cells and maroon are body cells. (**b**) Microlumens form; apical polarity established (red outline in **b**, and **c**). (**c**) Microlumens continue to expand during the primary transition, leading to the formation of a branching epithelium. (**d**) With the onset of the secondary transition, the tips (blue) and trunk (maroon) of the branching ducts now possess different cell fates. Within the trunk are cells that will differentiate into endocrine cells (purple cells). Branches also continue to form via the merging of microlumens. (**e**) Branches expand, and new tips are generated that will grow into a complex arbor structure. (**f**) The MPCs (lavender) express specific transcription factors. With the upregulation and downregulation of additional transcription factors, differentiation occurs from the trunk and tip cells into acinar and duct/endocrine fates, respectively

Sox9, *Nkx6.1*, *Hnf1β*, *Glis family Zinc Finger 3* (*Glis3*), *Hnf6*, *Hes family bHLH transcription factor A1* (*Hes1*), *Ngn3*, and *Pdx1* (Solar et al. 2009). The expression of many of these genes is maintained in differentiated duct cells. Tip cells, on the other hand, differentiate into the acinar cell lineage and express *Myc proto-oncogene* (*c-Myc*), *Carboxypeptidase A1* (*Cpa1*), *amylase* (*Amy*), *Recombination signaling protein for Immunoglobulin Kappa J* (*RBP-J*), and *Ptf1a* (Zhou et al. 2007; Fujikura et al. 2006; Nakhai et al. 2008). Significant research has identified roles for all these genes/proteins in duct cell identity and development.

Two of the most well-established transcription factors expressed in the ductal epithelium early in the primary transition and maintained in the trunk duct cells are SOX9 and HNF1β. Despite the expression of both transcription factors in the trunk cells, their expression pattern is heterogeneous. SOX9$^+$/HNF1β$^-$ cells demark multipotent progenitors at the periphery between tip and trunk cells, while SOX9$^+$/HNF1β$^+$ cells have a greater propensity for a duct fate. The SOX9$^+$/HNF1β$^-$ cells also express NKX6.1 and PDX1, indicating their capacity for multipotency (Kopp et al. 2011). Interestingly, immunofluorescent imaging and fluorescence-activated cell sorting (FACS) analysis of adult mouse pancreas cells demonstrated consistent and homogenous HNF1β staining in large ducts, while SOX9 staining appeared stochastic (Rezanejad et al. 2018). Conversely, in small ductal termini, the opposite expression pattern was observed, with some cells expressing HNFβ while most expressed SOX9 robustly (Rezanejad et al. 2018). This heterogeneity between HNFβ and SOX9 may reflect a currently unappreciated role for these factors in the maintenance and/or function of mature duct subpopulations.

GLIS3 is also expressed in duct cells and regulates duct development. A *Glis3* knockout mouse model displayed pancreatic ducts that expanded into large cystic structures (Kang et al. 2016), a similar phenotype to that observed in the absence of HNF1β or HNF6. Specifically, *Hnf6*$^{-/-}$ mice had aberrant branching and formed large cystic structures in the interlobular and intralobular ducts but not the intercalated ducts (Jacquemin et al. 2000). *Hnf1β*$^{-/-}$ mice also developed cystic structures, and the authors noted decreased proliferation in the pancreata overall, suggesting that failed duct formation impairs the expansion of all cell types (De Vas et al. 2015). Furthermore, *Ngn3* expression and acinar cell identity were altered in *Hnf1β*$^{-/-}$ mice, indicating a role for HNF1β-expressing ductal cells in both endocrine progenitor cell development and acinar cell differentiation (De Vas et al. 2015). The similarities between the *Glis3*, *Hnf1β,* and *Hnf6* mutant pancreata indicate a collective role for these transcription factors in specifying ductal cells.

In addition to many specific transcription factors expressed in the ducts, signaling pathways also play a role in duct development. In particular, the notch signaling pathway has been extensively studied in the setting of pancreas development (Reviewed in (Li et al. 2015)). Notch is a single-pass transmembrane receptor that is encoded as four different isoforms, Notch1–4. Notch ligands located on the cell surface signal to adjacent cells via interaction with the extracellular domain of the Notch receptor. Ligand binding can then activate or inhibit proteolytic cleavage of the Notch intracellular domain (NICD). Translocation of the NICD to the nucleus results in transcriptional activation of RBP-J, which stimulates the *Hairy/Enhancer of split* (*Hes*) genes.

Previous work established a role for Notch in duct identity. Specifically, disrupting Notch signaling by targeted loss of *Rbp-j* in *Pdx1*-expressing cells (*Pdx1-Cre;Rbp-j*$^{loxP/loxP}$) resulted in a loss of branched structures at E15 with adult animals displaying an overall smaller pancreas and loss of β-cell mass at 10 weeks of age (Fujikura et al. 2006). More recent evidence suggests that oscillations in expression of Notch and timed depletion of Notch prior to E13 impact cell fate. Seymour et al. (Seymour et al. 2020) utilized a mouse model wherein a dominant-negative *Mastermind-like 1* (*Maml1*), normally a co-activator of HES1, was expressed in *Hnf1β* expressing cells (*Hnf1b-Cre*^{ERT2}*;Rosa26*^{dnMaml1}). This MAML inhibition resulted in a conversion of the bipotent ductal trunk cells to acinar-specified tip cells before E13. Additionally, modulating the expression of *Hes1* using either an inhibitor or activator of the NICD also altered cell-fate decisions via altered expression of SOX9 and PTF1a. Taken together, these studies demonstrate how regulation of bipotent trunk cell identity via Notch signaling is essential for pancreatic duct development.

These studies described some of the factors that contribute to duct cell identity in the developing and adult pancreas. It is increasingly clear that correctly differentiated duct cells are necessary for subsequent specification of the endocrine and acinar cell lineages. Duct development also requires a number of dynamic processes that permit the formation and structure of the ductal epithelium. This topic will be discussed in the following section.

1.2 Structural Organization of the Pancreatic Epithelium

1.2.1 Cap and Body Cells

The concept of tip and trunk cells illuminates the spatial organization of the progenitor cells and the differentiation of subsequent pancreatic cell lineages. Our understanding of ductal epithelium formation and organization was further facilitated by the identification of the "cap" and "body" cells (Fig. 2a). This terminology describes the spatial positioning of the pancreatic progenitors within the early dorsal and ventral pancreatic buds—before ductal structures manifest. Within the early pancreatic buds, the progenitor cells around the periphery form a "cap," while the cells within the bud compose the "body." The initial observation of divergences between cap and body cells beyond their spatial position was first described in a study by Villasenor and colleagues (Villasenor et al. 2010). By virtue of their distinct local environments, the cap and body cells acquire differences in cell polarization that result in variable contributions to the morphogenetic arrangement of populations that differentiate from these cells.

Branching morphogenesis is a critical process driven by the generation of ductal structures; it enables organ growth and formation. Branching morphogenesis is an example of how the differences between the cap and body cells contribute to pancreas development based on mechanisms relating to duct formation. Villasenor et al. (Villasenor et al. 2010) made fundamental contributions to our understanding of branching morphogenesis in the pancreas. Using images of the developing pancreas at different embryonic stages, the authors characterized pancreas shape and structure over time. Across more than 50 pancreata, critical events were observed, including the initiation of tip protrusions from the epithelium at E12.5. Unexpectedly, the tips did not elongate and expand in length but rather maintained their size and increased in number. This significant observation refuted the previous dogma that branches form by elongation.

The question now becomes: how do branches form in the pancreas, and how does tip number expansion facilitate this process? Villasenor et al. (Villasenor et al. 2010) observed that the pancreatic epithelium undergoes a transient phase of stratification. Initially, the epithelium is a single layer of cells (E8.5); however, as the epithelium continues to expand, more layers become apparent (E9.5), and the cells in these

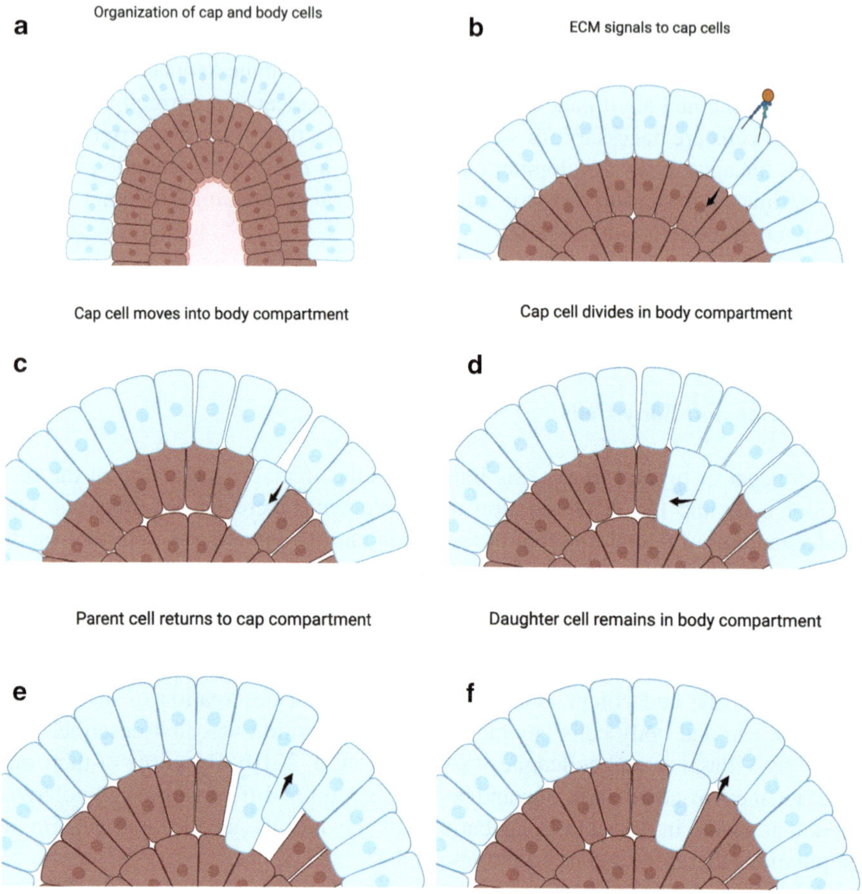

Fig. 2 Organization of the cap and body cell compartments. (**a**) Cap cells (blue) are located on the external layer of the pancreatic bud. Body cells (maroon) are found internal to the cap cells. Some body cells can line the internal pancreatic central lumen, and these body cells will develop apical polarity (pink). The extracellular matrix (ECM) surrounds the pancreatic bud and interacts with the cap cells due to their peripheral location. (**b**) Cap cells express integrins (green receptor with orange ligand) and other cell-surface proteins that interact with the ECM and function to initiate cellular movement, e.g., mitosis-associated cell dispersal (MACD). (**c**) Cap cells move into the body compartment (arrow). (**d**) Cap cell undergoes mitosis in the body compartment (arrow denotes generation of daughter cell). (**e**) After mitosis, the parent cell moves back to the cap compartment, while the daughter cell remains in the body compartment. (**f**) New cap cell in the body compartment

layers acquire and lose polarity with growth. By E10.5, the epithelium comprises at least five cell layers, and polarized cells begin to re-emerge. Subsequently, at E11.5, destratification occurs, resulting in the presence of only two or three cell layers.

Therefore, this process of stratification and destratification may control long branch development and generation of the ductal plexus through luminal network coalescence. Villasenor et al. (Villasenor et al. 2010) noted that as this ductal structure forms, the cap and body cells show polarity differences. Whereas the cap cells were polarized along their basal membrane due to contact with the extracellular matrix (ECM), the body cells were polarized along their apical membrane. This apical polarization resulted in body cells forming microlumens, the first instance of any duct-like structure in the pancreatic epithelium. Notably, these body cells eventually lost polarity during destratification to expand the ductal branches. Therefore, the formation of the pancreatic ductal structure relies on differences in polarization between cap and body cells to create luminal openings that will fuse and form branches.

As previously noted, the location of the cap cells at the periphery of the pancreatic bud positions the basal membrane of these cells in direct contact with the extracellular matrix, generating basal polarity. Therefore, how the ECM contributes to the dynamic behavior of the cap and body cells and its role in duct formation and branching morphogenesis must be addressed.

1.2.2 Extracellular Matrix

The ECM is a macromolecule network that provides essential structural support to tissues and mediates cell signaling via different growth factors and receptors. Collagen, proteoglycan, laminin, fibronectin, and elastin are proteins typically found in the ECM. Depending on the tissue and the specific context, additional growth factors and other molecules may be present in the ECM (Yue 2014). Cellular interactions with the ECM can promote proliferation, polarization, motility, and cell–cell adhesion. The ECM facilitates many cellular processes in the developing pancreatic buds.

The ECM most critically regulates motility and polarity in cap and body cells. The cap cells are organized along the periphery of the developing pancreatic bud (Fig. 2a). Resultantly, cap cells (not body cells) interact with the basement membrane, a specialized form of ECM (Fig. 2b). In addition to this association with the cap cells, other published studies have assisted our understanding of how the ECM contributes to ductal organization in the early pancreatic domain.

Work from Shih et al. (Shih et al. 2016) confirmed that early pancreatic organization is, in part, mediated by the motility difference between the cap and body cells (Fig. 2c–f). Time-lapse confocal imaging was used to observe a contrast in the dynamic behaviors between cap and body cells; specifically, cap cells were motile (i.e., capable of motion) and pleiomorphic (i.e., variable within their movement and final location) (Shih et al. 2016). Cap cells also exhibited a greater propensity for mitosis-associated cell dispersal (MACD). MACD is also observed in other branched epithelial structures, such as the ureteric bud of the kidney. MACD is the capacity for a cell to delaminate from the epithelium, undergo mitosis in the luminal region, and reinsert into the epithelium in a location distant from its division point (Packard et al. 2013). Cap cells frequently moved to the body compartment to undergo mitosis, after which the sister cell was dispersed back to the cap compartment. In contrast, body cells never underwent mitosis in the cap compartment.

These cellular movements contribute to the formation of the early pancreatic ductal structures.

Disrupting the interaction of the cap cells with the ECM inhibits ductal branching. Integrins are receptors on the cell surface that are commonly associated with ECM interactions (Kechagia et al. 2019). Shih and colleagues hypothesized that cap cells would receive signals from the ECM via integrins, and this transduction of information induced cap cell behaviors (Shih et al. 2016). A pancreas-specific integrin receptor knockout mouse model (*Pdx1-Cre*; *Itgb1*$^{loxP/loxP}$) specifically showed a severe absence of branching in the early pancreatic domain, which supports a role for integrin signaling in ductal morphogenesis (Shih et al. 2016). This finding demonstrates the influence of the ECM on ductal structure formation.

A recent study provides further evidence for the contribution of extrinsic mechanical cues via integrin signaling to pancreas development and cell fate decisions. Using human embryonic stem cells (hESCs), the authors determined that integrin signaling stimulates cell differentiation via mechanosensing (Mamidi et al. 2018). Mechanosensing refers to the capacity of cells or tissues to respond to mechanical cues (Kamkin and Kiseleva 2005). Yes1-associated transcription factor (YAP1) lies within the F-actin-Notch signaling cascade that responds to changes in mechanical signals by integrin α-5. Whereas YAP1 expression increased when cells were dispersed, YAP1 expression decreased when cells were more densely packed (Mamidi et al. 2018). Interestingly, dispersed cells differentiated into duct cells, whereas the more densely packed cells differentiated into endocrine cells. These results were confirmed in vivo with a *Yap1*$^{-/-}$ mouse model (Mamidi et al. 2018). Therefore, we can now understand how ECM interactions, which stimulate cell motility, might influence cell fate decisions by altering cell density between the cap and body compartments.

1.2.3 Ductal Lumen Formation

As discussed above, the ductal structure is critical for pancreatic development and differentiation. In addition to cellular and spatial interactions, lumen formation is a crucial event in the structural organization of the ductal epithelium. Lumenogenesis in the pancreas initiates from clusters of polarized cells that generate rosette-like structures around a central lumen. Lumenogenesis is dependent on the acquisition and maintenance of polarity by the body cells within the pancreatic bud. As outlined above, the MPCs in the early pancreatic buds are stratified into layers of cap and body cells. In addition to occupying specific spatial locations, the cap cells are polarized at their basal membrane, and the body cells show apical polarity (Villasenor et al. 2010). The innermost body cells acquire apical polarity and surround the central pancreatic lumen, which appears within each bud around E10 (Villasenor et al. 2010). The remaining body cells are unpolarized at this time. Interestingly, some body cells re-acquire polarity to form microlumens; this polarity reacquisition is essential for microlumen formation (Villasenor et al. 2010). As shown in Fig. 3,

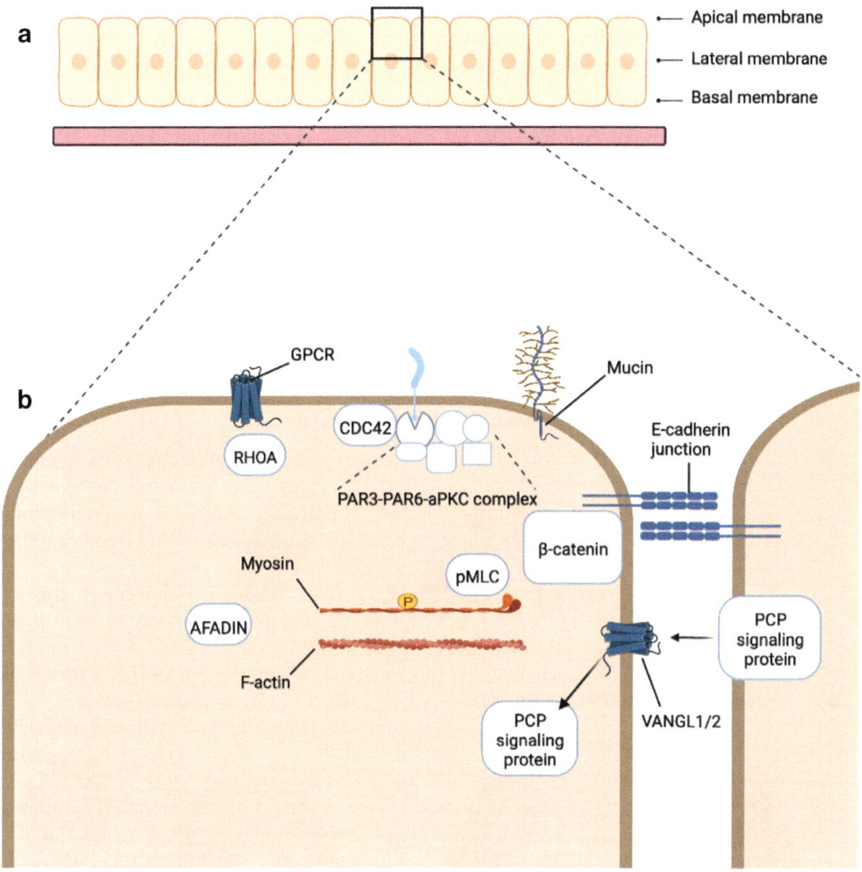

Fig. 3 Apico-basal and planar cell polarity. (**a**) A diagram illustrating the apical, lateral, and basal membranes. Boxed area is magnified in **b**. (**b**) A diagram illustrating factors that play a role in apico-basal polarity and planar cell polarity (PCP). *aPKC* atypical protein kinase C; *pMLC* phosphomyosin light chain complex; *VANGL* VANGL planar cell polarity 1

many genes and protein complexes contribute to the acquisition and maintenance of polarity, which is critical for lumen formation.

Cell division cycle 42 (*Cdc42*) is a member of the Rho-GTPase family and is part of a signaling pathway that establishes cell polarity in the pancreas (Etienne-Manneville 2004). The specific role of CDC42 in the developing pancreas was first explored in 2009 by Kesavan and colleagues (Kesavan et al. 2009). *Cdc42*$^{-/-}$ mice exhibit impaired initiation of lumen formation and failed maintenance of apical cell polarity due to the loss of other apical polarity markers, including Mucin1 (MUC1), Partitioning defective 6 (PAR6), and F-actin (Kesavan et al. 2009). The contribution of the Rho-GTPase protein family has been further investigated in recent years. Afadin and RhoA are two additional apical polarity factors involved in pancreatic lumen formation and maintenance. Afadin is an actin filament-binding protein

(Ikeda et al. 1999), and RhoA is a member of the Rho-GTPase family that functions in actin-filament regulation (Aspenström et al. 2004). Cooperation between these polarity factors is critical for generating a ductal network and permitting subsequent endocrine cell differentiation (Azizoglu et al. 2017). A mouse model with a pancreatic, genetic deletion of Afadin (*Pdx1-Cre;Afa*$^{loxP/loxP}$) demonstrated no loss of the initial pancreatic lumens; however, lumen continuity in the tip epithelium and de novo tip lumenogenesis were disrupted (Azizoglu et al. 2017). Figure 4 depicts the general process of lumen formation. Interestingly, the *Afa*$^{-/-}$ animals survived due to some redundancy with RhoA.

PDX1 plays an indispensable role in specifying pancreatic progenitor cells. Due to the essential nature of lumen formation in pancreas development, PDX1 likely

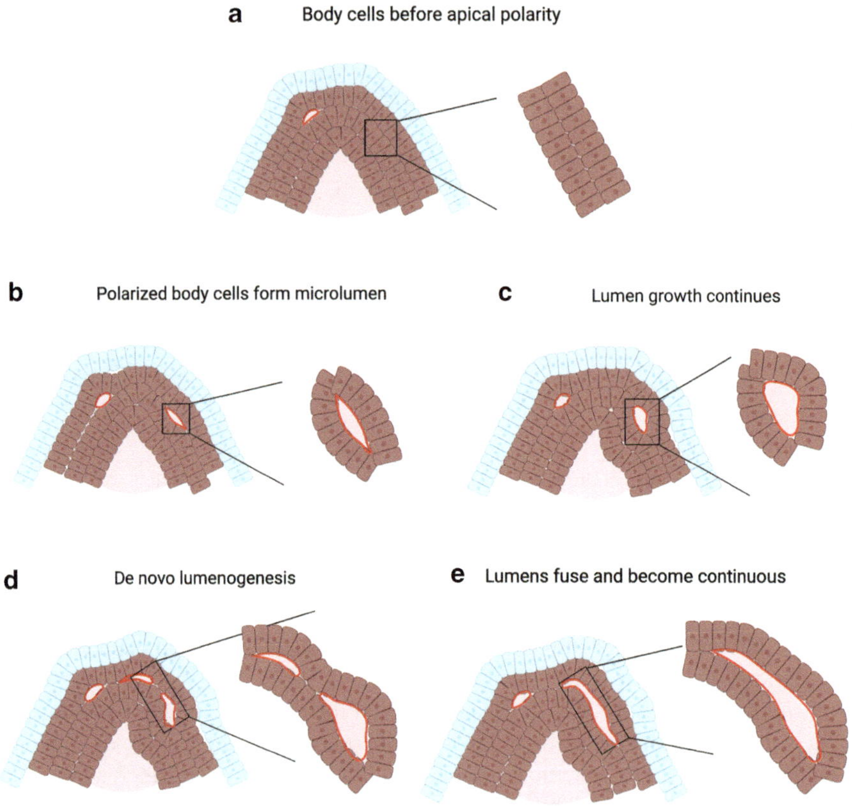

Fig. 4 Lumen structure, de novo lumenogenesis, and lumen continuity. (**a**) Early pancreatic bud, before extensive lumen formation. Cap cells surround the outside and body cells are internal. In all images, the black box designates the lumen of focus magnified on the right. (**b**) The development of new lumens (de novo lumens) and the generation of lumen continuity (connected lumens). The body cells become polarized and begin to develop an opening. (**c**) The lumen expands. (**d**) Additional microlumens develop. (**e**) Lumens can fuse to generate continuous ductal structures, which lead to the branching ductal tree in the pancreas

plays a role in regulating lumen formation. A study explored this possibility using a Pdx1 null (*Pdx1*$^{LacZ/LacZ}$) mouse model (Marty-Santos and Cleaver 2016). The pancreatic domain in this mouse was visualized using MUC1, a protein localized to the apical membrane of the pancreatic lumen-lining cells, and E-cadherin, a cell–cell adhesion protein (van Roy and Berx 2008) expressed on all epithelial cells. *Pdx1*$^{LacZ/LacZ}$ cells formed microlumens at E11.5. However, the lumen number was reduced compared with controls, decreasing further as development proceeded. Notably, this reduction was not due to a lack of microlumen formation. Instead, the microlumens that formed were unable to regulate their size and thus expanded into cysts (Marty-Santos and Cleaver 2016). Lumen diameter was noticeably increased as early as E11.5.

Furthermore, the authors noted that expression of E-cadherin and β-catenin, which form a complex at epithelial adherens junctions (MacDonald et al. 2009), was decreased at E11.5, illustrating dysfunctional cell–cell adhesion concomitant with microlumen formation/expansion. Finally, the phospho-myosin light chain complex (pMLC) is a protein complex involved in cellular adhesion and is expressed downstream of E-cadherin signaling. Activation of pMLC preserves apical constriction (Martin and Goldstein 2014). Apical constriction reduces the size of the apical end of the cell compared with the basal end, resulting in a more pyramidal cell shape. From E11.5, pMLC expression at the apical surface was markedly decreased, contributing to the failure of apical polarity and lumen integrity maintenance (Marty-Santos and Cleaver 2016). Ultimately, disrupting these processes resulted in decreased MPC number and proliferation (Marty-Santos and Cleaver 2016). This study emphasizes how the cellular processes relevant to lumen formation contribute to overall organ structure and development.

Planar cell polarity (PCP) is another pathway that contributes to the organization of epithelial sheets and, as such, has been implicated in pancreatic duct formation. Figure 3 illustrates apicobasal polarity and PCP. Although the role of apicobasal polarity has been well described in the context of pancreatic duct development (see above), the role of the PCP pathway is not as well defined; in fact, only select factors in the PCP pathway have been studied in the pancreas. In particular, VANGL planar cell polarity proteins 1 and 2 (Vangl1, Vangl2) are transmembrane proteins that are central components of PCP signaling (Butler and Wallingford 2017). The presence of PCP proteins in the pancreas was first established in 2012 by Cortijo and colleagues (Cortijo et al. 2012). The authors identified the expression of *Vangl1* in Pdx1^{+} cells as early as E10.5 and subsequent restriction of *Vangl1* to cells of the ductal epithelium but exclusion from differentiating endocrine cells (Cortijo et al. 2012). In 2020, Flasse and colleagues (Flasse et al. 2020) demonstrated that Vangl1/2 expression was localized to the apical side of the lateral membrane of cells in the pancreatic ducts. Significantly, when Vangl2 localization was disrupted at E12.5, a significant loss of pancreatic progenitor cells and a disruption of epithelial integrity occurred. Critically, the absence of PCP signaling due to altered Vangl2 disrupted lumen formation (Flasse et al. 2020).

Taken together, these studies illustrate the essential role of cell polarity in ductal lumen formation and the structure of the pancreatic epithelium. Disrupting either apicobasal polarity or PCP signaling results in a failure to properly generate ducts, directly impacting subsequent cellular differentiation. In the next section, we review emerging tools for researching the dynamics of pancreatic ductal development and organization.

1.3 Tools for Understanding Duct Development and Organization

The study of organ development, and in this case, the specific exploration of pancreatic duct development and organization, is greatly assisted by tool creation. The recent generation of novel cellular lineage tracing and modeling tools has provided insight into how the pancreatic ducts develop and are organized.

1.3.1 Cellular Lineage Tracing

Lineage tracing has been an instrumental tool for understanding how progenitor cells contribute to the differentiation and maturation of various cell populations within an organ. In fact, many of the studies outlined above used lineage tracing to understand specific phenotypic observations. An example is a study by Zhou et al. (Zhou et al. 2007), which used lineage tracing to establish the model of tip and trunk cells. Therefore, the progression to more sophisticated or accurate methods would significantly assist in the visualization of certain aspects of development and the generation of new hypotheses regarding dynamic changes in the pancreas during development.

Clonal lineage tracing is the most common method used. Studies like those of Zhou et al. (Zhou et al. 2007) used a single color fluorescent reporter that permanently marked cells expressing a specific protein, and all of their descendants. Sznurkowska and colleagues (Sznurkowska et al. 2018) used a more complex model (*Rosa26*-Cre^{ERT2}/*R26R*^{Confetti}), wherein administration of low-dose tamoxifen to pregnant females activated Cre recombinase, which then induced clonal labeling. However, a significant difference in this model compared with other lineage trace reporters was the "Confetti reporter," which allows populations of clonal cells to be labeled with a unique color. Labeling more than one clonal population of cells permits the examination of the fate of many distinct cells, and their descendants, in the early developing pancreas. Using this visualization tool, the authors discovered that lineage specification occurs earlier than previously appreciated (Sznurkowska et al. 2018). For example, clones traced from E12.5 to postnatal day 14 were found to differentiate into only one cell type in the pancreas.

This lineage tracing study also revealed that most ductal branches were clonal, i.e., derived from a single cell (Sznurkowska et al. 2018). Additionally, the acinar cells proximal to the ductal branches were derived from their closest ductal clone

(Sznurkowska et al. 2018). These observations indicated that cells at or near the tip of the expanding ductal epithelium self-renew and expand continuously to maintain and expand the ductal epithelium. They also generate acinar cells at the tips through rounds of serial branching (Sznurkowska et al. 2018). Moreover, the observation that branched structures were either completely labeled or completely unlabeled with a single color indicated that ductal subtrees expand from a small number of tip progenitor cells. Therefore, this study discovered that tip progenitor cells are not only a niche for acinar differentiation, as previously observed, but also a niche for a population of self-renewing ductal cells (Sznurkowska et al. 2018). Moreover, this study demonstrates the power of developmental biology tools to facilitate a deeper understanding of ductal cell behavior and differentiation and identify aspects of pancreas development not previously appreciated.

1.3.2 Modeling

The termination of the primary transition at E12.5 initiates the secondary transition. This stage is the developmental period in which the specified tip and trunk cells undergo cellular differentiation to their mature cell types. How the ductal epithelium remodels as the pancreas progresses from the primary to secondary transition and during the secondary transition is incompletely understood (Fig. 1). Studying this remodeling process is hindered by a lack of tools to visualize the ducts within an embryo *in utero*. Advances in *in silico* modeling now provide a means to observe the dynamic processes involved in forming the pancreatic ductal networks during development. In fact, *in silico* modeling has been used to define the principles regulating ductal network formation in the pancreas (Dahl-Jensen et al. 2018). Specifically, whole-mount immunofluorescent images of the pancreas showing expression of the ductal epithelial markers Mucin1 and E-cadherin were generated and digitized. These digital images provided an outline for the pancreatic epithelium at three stages during development: E12.5 (end of the primary transition), E14.5 (middle of the secondary transition), and E18.5 (end of pancreas development *in utero*). Changes in the ductal structure at these ages were then used to make computer modeling-based predictions concerning ductal network formation. A significant observation from this modeling was that the transition from a lobular plexus at E12.5 to an arbor at E18.5 was dependent on the distance from an acinus to the duodenum (Dahl-Jensen et al. 2018). In other words, during the branching process, the branching of a duct is influenced to form the shortest length possible between the termini containing acinar cells and the duodenum. A short distance between the acini and the duodenum is beneficial since the duodenum is the exit point for pancreatic fluid and thus has the highest fluid flow throughout the ducts. Further experimentation using animal or organoid models is required to confirm these observations; however, *in silico* modeling has provided valuable insight that has resulted in new hypotheses regarding pancreatic ductal system development.

In silico modeling has also provided a quantitative description of ductal development. Transportation networks, rivers, and blood vessels can all be modeled using fluid dynamic principles. Given the structural resemblance of the pancreatic ducts to vasculature, Dahl-Jensen and colleagues (Dahl-Jensen et al. 2018) employed these principles to understand duct development. Using images of E14.5 pancreata, a model depicting pancreatic duct stimulation of fluid flow (i.e., flux) was generated. Modeling flux clarified the mechanical signals required for the process of branch remodeling. This model specified the terminal ends where the acinar cells are located as the point where fluid enters the main pancreatic duct; this main pancreatic duct is connected to the duodenum, which is the point where fluid exits. The model showed that the greatest flux occurs near the connection of the main pancreatic duct to the duodenum, which was expected given that the main duct is the widest and is the convergence point for all pancreatic fluid. Additional analyses discovered that when redundant ducts were generated, i.e., ducts that could be eliminated without fragmenting the flow of fluid to the main duct, redundant ducts with the greatest flux (the most flow of fluid) would not be pruned, but those with low flux would be eliminated. To understand these processes, two different models of pruning were generated, one in which branches were randomly pruned regardless of flux, and a second in which flux dictated branch pruning. Dahl-Jensen et al. (Dahl-Jensen et al. 2018) then compared the resulting networks with those experimentally derived from the E14.5 digitized imaging data. The computer-generated network with "pruning-based flux" resulted in branches pruned from the periphery toward the center, i.e., from the lowest flow to the greatest flow. The structure of the "flux-based pruning" network also resembled the experimentally derived data. Therefore, this study provides an excellent example for the use of computer-based modeling to assist our understanding of network formation in the pancreas and the contribution of ductal fluid flow to the ductal tree structure (Dahl-Jensen et al. 2018).

Computer modeling tools provide an opportunity to generate new theories and hypotheses that can then be experimentally validated and tested. The following section discusses the contribution of ductal disorganization and dysfunction to disease. The use of *in silico* modeling may be a valuable tool to assist future studies of disease pathogenesis and progression.

1.4 The Role of the Pancreatic Ducts in Organ Function and Dysfunction

1.4.1 The Ductal System Is Essential for Pancreatic Function

As in most biological systems, the structure and function of the pancreatic ducts are closely related. The epithelial cells that form the pancreatic ducts act as a funnel to transport proenzymes to the common bile duct, which originates at the liver. These ducts meet to form the Ampulla of Vater, where enzymes from the liver and pancreas mix and are delivered to the duodenum upon stimulation, i.e., feeding (Pandol 2010). The acinar cells of the exocrine pancreas are responsible for the synthesis and subsequent secretion of the digestive proenzymes (also known as zymogens) into the ducts. These proenzymes must remain inactive until their function is required for nutrient digestion. To maintain an inactive state, ductal cells secrete bicarbonate into the pancreatic ducts, where it acts as a chemical buffer allowing for the safe transport of catalytically inactive proenzymes such as trypsinogen, chymotrypisnogen, procarboxypeptidase, and proelastase. Once in the duodenum, the proenzymes are activated by either enteropeptidase, which is located in the brush border membrane, or by autocatalytic activation (Pandol 2010). This process occurs most readily in the presence of bile salts, which proenzymes normally do not encounter until immediately before entering the duodenum (Sarkany and Moreland 1985). Once activated, these enzymes break down nutrients in the duodenal chyme, making them bioavailable and readily absorbed.

Trypsin, in particular, plays a significant role in autocatalysis and the catalysis of other enzymes present in the pancreatic juice. Therefore, it is crucial that this particular digestive enzyme remains inactive within the acinar cell granules and pancreatic ducts. As an additional safeguard, the ductal lumens of the intestine and pancreas are protected from the abrasive traffic by a layer of alkaline mucus and cilia, which act as street sweepers to move the ductal contents. Unsurprisingly, diseases that lead to obstruction of the pancreatic ducts or thickening of its mucus directly endanger the exocrine pancreas, and indirectly the islets, as well as the function of the organ as a whole. Blocking the flow of proenzymes/zymogens can result in inappropriately timed activation, leading to pancreatic inflammation and possible autodigestion. In this disease state, supplementation with digestive enzymes (a therapeutic replacement for exocrine function) or insulin (a therapeutic replacement for endocrine function) may reverse some of the dysfunction and return the organ to proper function. Clearly, the pancreatic ducts are essential for normal pancreatic function, a fact best illustrated by examples of pancreatic pathology caused by ductal dysfunction.

1.4.2 Cystic Fibrosis

The disease cystic fibrosis (CF) results from loss-of-function mutations in the CF transmembrane conductance receptor (*CFTR*), which leads to a dysfunction in ion trafficking and ultimately viscous epithelial mucin secretions in any organ where CFTR is expressed, including the lungs and pancreas (Madácsy et al. 2018). CFTR is a key player in the secretion of bicarbonate from pancreatic ductal cells. As mentioned above, bicarbonate functions to buffer pancreatic juice and acidic chyme, facilitating optimal activity of the digestive enzymes in the duodenum. Pancreatic ductal cell dysfunction results in an increase in pancreatic juice acidity, which reduces the overall secretory volume and leads to increased protein content in the ductal fluid. These changes contribute significantly to the pancreatic pathology observed in CF (Kopelman et al. 1988, 1985; Tucker et al. 2003; Wilschanski and Novak 2013).

The initial insult to the pancreas in CF is an accumulation of zymogens and mucus in the ducts due to dysfunctional CFTR (Tucker et al. 2003). In the pancreas,

CFTR exists exclusively on the apical (luminal) membrane of the ductal epithelium. As a result, CFTR directly participates in the secretion of bicarbonate and water into the ducts; mutations in CFTR interrupt this function. There are six classes of CF-causing mutations, five of which result in dysfunctional or absent CFTR (85% of patients). In these patients, pancreatic insufficiency is clinically observed, indicating that dysfunction of ductal secretion contributes to exocrine loss (Wilschanski and Novak 2013). However, those patients that remain pancreatic sufficient can present with pancreatitis, which is a major risk factor for developing diabetes. Only 50% of adults with CF develop CF-related diabetes (CFRD); in all cases, exocrine dysfunction precedes the loss of islet cell mass and/or β-cell dysfunction (Moran et al. 2009; Barrio 2015). This observation is compelling evidence that the exocrine pancreas, and not the islets, is the primarily affected compartment of the organ (Hart et al. 2016). Once dysfunctional CFTR hinders the ability for enzymes to be flushed from the ductal lumen, the exocrine pancreas becomes damaged and undergoes autolysis. The resultant production of inflammatory cytokines stimulates nearby immune cells, leading to inflammation of the whole organ and (in some cases) inhibition of islet secretory function (Hart et al. 2016; Donath et al. 2010). In a study by Hart et al. (Hart et al. 2018), exocrine tissue and islets were significantly reduced in pancreata from their CF model ($MIPCre^{ERT};Cftr^{loxP}/Cftr^{\Delta 11}$); however, isolated islets maintained secretory function. This result strongly indicates that CF does not directly impact islet function, but rather the loss of proper ductal function has secondary effects on both the exocrine and endocrine cells. As clearly evidenced by CF pathology, proper pancreatic ductal function is critical for organ function.

1.4.3 Type 3c Diabetes

Type 3c diabetes (T3cD), also referred to as "pancreatogenic diabetes," results from damage to the acinar cells and ducts that subsequently negatively impacts endocrine function, leading to hyperglycemia. The most common etiology of T3cD begins with sustained acinar disease leading to ductal obstruction, acinar inflammation, and fibro-fatty replacement of the exocrine pancreas (Rickels et al. 2020). The spectrum of clinical phenotypes observed with T3cD directly relates to how the sustained acinar disease arises. However, in general, without a fully functioning exocrine pancreas, the islets become damaged, which leads to their dysfunction. The causes of T3cD, by definition, must involve dysfunction of the pancreatic ducts at some point in the disease process. This means that ductal dysfunction could either initiate the organ damage (such as in CF) or become dysfunctional as a result of a physical blockage (such as with gallstones or tumorous growth). Similar to acute and chronic pancreatitis, a specific and effective therapeutic target has yet to be identified that could directly target the initiating event(s) within the pancreas to reverse the clinical pathology.

1.4.4 Ciliopathies

Primary cilia are important organelles that play a role in cell migration, fluid movement, and sensory reception. "Ciliopathies" are diseases that result from the dysfunction or absence of the primary cilia. Polycystic kidney disease (PKD) is one example of a ciliopathy with primary cilia dysfunction, leading to uncontrolled epithelial growth and proliferation as a result of ciliary-dependent signaling, usually in the distal nephron and/or collecting duct (Bergmann et al. 2018; McConnachie et al. 2021). Interest in the role of cilia as it relates to pancreatic homeostasis came from observations that patients with PKD also sometimes developed pancreatic ductal cysts.

Several studies have utilized induced mutations in ciliary genes to study the involvement of cilia in disease pathogenesis. Zhang et al. (Zhang et al. 2005) investigated the pancreatic pathology caused by a transgene insertion-induced hypomorphic allele, *Tg737* (*Tg737*$^{delta2\text{-}3betagal}$). Tg737 encodes an intraflagellar transport protein (*Ift88*) essential for cilia formation (Zhang et al. 2005). Although multiple organs are affected in this model, the pancreatic phenotype directly demonstrates the importance of the ducts in the disease etiology. Despite the discovery that cilia were expressed only on duct and islet cells (not acinar cells), the phenotype showed significant acinar and ductal dysfunction. In particular, the mutant pancreata showed severe acinar cell apoptosis, acinar cell vacuolization, increased amylase release, abnormal activation of intrapancreatic carboxypeptidase, increased proliferation of the ductal epithelium, and cystic structures in the ducts. This mouse model suggested that the absence of cilia hindered ion transport in the ductal epithelium, resulting in premature activation of digestive enzymes (as seen is CF). Further study is required to understand the signaling events that occur between the ducts and acini to support cellular survival and growth (Zhang et al. 2005).

Further evidence for the role of cilia in pancreatic duct development and function was provided by additional mutant mouse models. Specifically, the pancreata analyzed from mouse models with genetic deletion of *Hnf6* or Liver Kinase B1/ Serine/Threonine Kinase 11 (*Lkb1*), wherein either gene was postnatally inactivated using an inducible Cre driven from the *Sox9* promoter (*Sox9-Cre*^{ER};*Hnf6*$^{loxP/-}$ and *Sox9-Cre*^{ER};*Lkb1*$^{loxP/loxP}$, respectively), showed ductal dilation as well as characteristics associated with chronic pancreatitis (acinar apoptosis, inflammatory infiltrates, fibrosis, and lipomatosis). These results suggest that ductal ciliary defects perturb acinar homeostasis and differentiation (Augereau et al. 2016). Taken together, these models reveal that ductal cilia are critical to maintain acinar stability during development and in the postnatal pancreas. Clearly, ductal dysfunction induced in a multitude of ways can impact the health and function of the pancreas.

2 Conclusion

In this chapter, we have described how the ductal cells are specified, form the pancreatic epithelium, and subsequently differentiate into duct, endocrine, and acinar cells. The specification of the pancreas MPCs requires the co-expression of *Pdx1* and *Ptf1a*, which establish the genetic regulatory networks necessary to instruct the differentiation of all pancreatic lineages. The ductal cells differentiate from the MPCs and are characterized by the expression of a specific set of transcription factors, depending on their location within the pancreatic epithelium. Ductal cells in the "trunk" can express *Sox9, Hnf1β, Pdx1, Hnf6, Nkx6.1, Glis3, Hes1,* and *Ngn3* and have the potential to differentiate into endocrine cells or remain as duct cells. The continued expression of *Hnf1β* and *Sox9* is maintained in mature ductal cells. The "tip" cells differentiate into acinar cells and maintain expression of *C-Myc, Cpa1, Amy, RBP-J,* and *Ptf1a*. However, before cellular differentiation from the ducts can occur, the branching ductal structure within the pancreas must be created. This begins in the earliest pancreatic buds with the organization and re-organization of the "cap" and "body" cells. The designation of cap or body cell within the initial pancreatic bud stems from their spatial location in the bud. The cap cells form at the periphery of the bud, while the body cells are internal. Organization and re-organization of the cap and body cells occurs due to changes in cell polarity and interactions with the ECM. This induces cell movement as well as various signaling pathways that lead to lumen formation, fusion, elongation, and branching. The use of animal and computer-based developmental biology tools continues to assist our understanding of pancreatic ductal development, which also provides the ability to understand how the pancreatic ducts dysfunction in the setting of disease.Compliance with Ethical StandardsDorian J. Dale of declares no conflict of interest. Caleb D. Rutan declares no conflict of interest. Teresa L. Mastracci declares no conflict of interest. This chapter does not contain any studies with human participants or animals performed by any of the authors.

References

Aspenström P, Fransson A, Saras J (2004) Rho GTPases have diverse effects on the organization of the actin filament system. Biochem J 377(Pt 2):327–337

Augereau C, Collet L, Vargiu P, Guerra C, Ortega S, Lemaigre FP et al (2016) Chronic pancreatitis and lipomatosis are associated with defective function of ciliary genes in pancreatic ductal cells. Hum Mol Genet 25(22):5017–5026

Azizoglu DB, Braitsch C, Marciano DK, Cleaver O (2017) Afadin and RhoA control pancreatic endocrine mass via lumen morphogenesis. Genes Dev 31(23–24):2376–2390

Barrio R (2015) Management of endocrine disease: cystic fibrosis-related diabetes: novel pathogenic insights opening new therapeutic avenues. Eur J Endocrinol 172(4):R131–R141

Bastidas-Ponce A, Scheibner K, Lickert H, Bakhti M (2017) Cellular and molecular mechanisms coordinating pancreas development. Development 144(16):2873–2888

Bergmann C, Guay-Woodford LM, Harris PC, Horie S, Peters DJM, Torres VE (2018) Polycystic kidney disease. Nat Rev Dis Primers 4(1):1–24
Bort R, Martinez-Barbera JP, Beddington RSP, Zaret KS (2004) Hex homeobox gene-dependent tissue positioning is required for organogenesis of the ventral pancreas. Development 131(4):797–806
Burlison JS, Long Q, Fujitani Y, Wright CVE, Magnuson MA (2008) Pdx-1 and Ptf1a concurrently determine fate specification of pancreatic multipotent progenitor cells. Dev Biol 316(1):74–86
Butler MT, Wallingford JB (2017) Planar cell polarity in development and disease. Nat Rev Mol Cell Biol 18(6):375–388
Cortijo C, Gouzi M, Tissir F, Grapin-Botton A (2012) Planar cell polarity controls pancreatic beta cell differentiation and glucose homeostasis. Cell Rep 2(6):1593–1606
Dahl-Jensen SB, Yennek S, Flasse L, Larsen HL, Sever D, Karremore G et al (2018) Deconstructing the principles of ductal network formation in the pancreas. PLoS Biol 16(7):e2002842
De Vas MG, Kopp JL, Heliot C, Sander M, Cereghini S, Haumaitre C (2015) Hnf1b controls pancreas morphogenesis and the generation of Ngn3+ endocrine progenitors. Development 142(5):871–882
Deutsch G, Jung J, Zheng M, Lóra J, Zaret KS (2001) A bipotential precursor population for pancreas and liver within the embryonic endoderm. Development 128(6):871–881
Domínguez-Muñoz JE (2011) Pancreatic exocrine insufficiency: diagnosis and treatment. J Gastroenterol Hepatol 26(s2):12–16
Donath MY, Böni-Schnetzler M, Ellingsgaard H, Halban PA, Ehses JA (2010) Cytokine production by islets in health and diabetes: cellular origin, regulation and function. Trends Endocrinol Metab 21(5):261–267
Ehre C, Ridley C, Thornton DJ (2014) Cystic fibrosis: an inherited disease affecting mucin-producing organs. Int J Biochem Cell Biol 52:136–145
Etienne-Manneville S (2004) Cdc42—the centre of polarity. J Cell Sci 117(Pt 8):1291–1300
Flasse L, Yennek S, Cortijo C, Barandiaran IS, Kraus MRC, Grapin-Botton A (2020) Apical restriction of the planar cell polarity component VANGL in pancreatic ducts is required to maintain epithelial integrity. Cell Rep 31(8):107677
Fujikura J, Hosoda K, Iwakura H, Tomita T, Noguchi M, Masuzaki H et al (2006) Notch/Rbp-j signaling prevents premature endocrine and ductal cell differentiation in the pancreas. Cell Metab 3(1):59–65
Hart PA, Bellin MD, Andersen DK, Bradley D, Cruz-Monserrate Z, Forsmark CE et al (2016) Type 3c (pancreatogenic) diabetes mellitus secondary to chronic pancreatitis and pancreatic cancer. Lancet Gastroenterol Hepatol 1(3):226–237
Hart NJ, Aramandla R, Poffenberger G, Fayolle C, Thames AH, Bautista A et al (2018) Cystic fibrosis-related diabetes is caused by islet loss and inflammation. JCI Insight 3(8):98240
Haumaitre C, Barbacci E, Jenny M, Ott MO, Gradwohl G, Cereghini S (2005) Lack of TCF2/vHNF1 in mice leads to pancreas agenesis. PNAS 102(5):1490–1495
Hebrok M, Kim SK, Melton DA (1998) Notochord repression of endodermal sonic hedgehog permits pancreas development. Genes Dev 12(11):1705–1713
Hebrok M, Kim SK, St Jacques B, McMahon AP, Melton DA (2000) Regulation of pancreas development by hedgehog signaling. Development 127(22):4905–4913
Hegyi E, Sahin-Tóth M (2017) Genetic risk in chronic pancreatitis: the trypsin-dependent pathway. Dig Dis Sci 62(7):1692–1701
Ikeda W, Nakanishi H, Miyoshi J, Mandai K, Ishizaki H, Tanaka M et al (1999) Afadin: a key molecule essential for structural organization of cell-cell junctions of polarized epithelia during embryogenesis. J Cell Biol 146(5):1117–1132
Jacquemin P, Durviaux SM, Jensen J, Godfraind C, Gradwohl G, Guillemot F et al (2000) Transcription factor hepatocyte nuclear factor 6 regulates pancreatic endocrine cell differentiation and controls expression of the proendocrine gene ngn3. Mol Cell Biol 20(12):4445–4454
Jonsson J, Carlsson L, Edlund T, Edlund H (1994) Insulin-promoter-factor 1 is required for pancreas development in mice. Nature 371(6498):606–609

Kamkin A, Kiseleva I (2005) Mechanosensitivity of cells from various tissues. In: Kamkin A, Kiseleva I (eds) Mechanosensitivity in cells and tissues. Academia, Moscow
Kang HS, Takeda Y, Jeon K, Jetten AM (2016) The spatiotemporal pattern of Glis3 expression indicates a regulatory function in bipotent and endocrine progenitors during early pancreatic development and in beta, PP and ductal cells. PLoS One 11(6):e0157138
Kawaguchi Y, Cooper B, Gannon M, Ray M, MacDonald RJ, Wright CVE (2002) The role of the transcriptional regulator Ptf1a in converting intestinal to pancreatic progenitors. Nat Genet 32(1):128–134
Kechagia JZ, Ivaska J, Roca-Cusachs P (2019) Integrins as biomechanical sensors of the microenvironment. Nat Rev Mol Cell Biol 20(8):457–473
Kesavan G, Sand FW, Greiner TU, Johansson JK, Kobberup S, Wu X et al (2009) Cdc42-mediated tubulogenesis controls cell specification. Cell 139(4):791–801
Kopelman H, Durie P, Gaskin K, Weizman Z, Forstner G (1985) Pancreatic fluid secretion and protein hyperconcentration in cystic fibrosis. N Engl J Med 312(6):329–334
Kopelman H, Corey M, Gaskin K, Durie P, Weizman Z, Forstner G (1988) Impaired chloride secretion, as well as bicarbonate secretion, underlies the fluid secretory defect in the cystic fibrosis pancreas. Gastroenterology 95(2):349–355
Kopp JL, Dubois CL, Schaffer AE, Hao E, Shih HP, Seymour PA et al (2011) Sox9+ ductal cells are multipotent progenitors throughout development but do not produce new endocrine cells in the normal or injured adult pancreas. Development 138(4):653–665
Krapp A, Knöfler M, Ledermann B, Bürki K, Berney C, Zoerkler N et al (1998) The bHLH protein PTF1-p48 is essential for the formation of the exocrine and the correct spatial organization of the endocrine pancreas. Genes Dev 12(23):3752–3763
Larsen HL, Grapin-Botton A (2017) The molecular and morphogenetic basis of pancreas organogenesis. Semin Cell Dev Biol 66:51–68
Li XY, Zhai WJ, Teng CB (2015) Notch signaling in pancreatic development. Int J Mol Sci 17(1):48
MacDonald BT, Tamai K, He X (2009) Wnt/β-catenin signaling: components, mechanisms, and diseases. Dev Cell 17(1):9–26
Madácsy T, Pallagi P, Maleth J (2018) Cystic fibrosis of the pancreas: the role of CFTR channel in the regulation of intracellular Ca2+ signaling and mitochondrial function in the exocrine pancreas. Front Physiol 9:1585
Mamidi A, Prawiro C, Seymour PA, de Lichtenberg KH, Jackson A, Serup P et al (2018) Mechanosignalling via integrins directs fate decisions of pancreatic progenitors. Nature 564(7734):114–118
Martin AC, Goldstein B (2014) Apical constriction: themes and variations on a cellular mechanism driving morphogenesis. Development 141(10):1987–1998
Marty-Santos L, Cleaver O (2016) Pdx1 regulates pancreas tubulogenesis and E-cadherin expression. Development 143(1):101–112
McConnachie DJ, Stow JL, Mallett AJ (2021) Ciliopathies and the kidney: a review. Am J Kidney Dis 77(3):410–419
Moran A, Dunitz J, Nathan B, Saeed A, Holme B, Thomas W (2009) Cystic fibrosis–related diabetes: current trends in prevalence, incidence, and mortality. Diabetes Care 32(9):1626–1631
Nakhai H, Siveke JT, Mendoza-Torres L, Schmid RM (2008) Conditional inactivation of Myc impairs development of the exocrine pancreas. Development 135(19):3191–3196
Packard A, Georgas K, Michos O, Riccio P, Cebrian C, Combes AN et al (2013) Luminal mitosis drives epithelial cell dispersal within the branching ureteric bud. Dev Cell 27(3):319–330
Pan FC, Bankaitis ED, Boyer D, Xu X, Van de Casteele M, Magnuson MA et al (2013) Spatiotemporal patterns of multipotentiality in Ptf1a-expressing cells during pancreas organogenesis and injury-induced facultative restoration. Development 140(4):751–764
Pandol SJ (2010) The exocrine pancreas. Morgan & Claypool Life Sciences, San Rafael
Rezanejad H, Ouziel-Yahalom L, Keyzer CA, Sullivan BA, Hollister-Lock J, Li WC et al (2018) Heterogeneity of SOX9 and HNF1β in pancreatic ducts is dynamic. Stem Cell Rep 10(3):725–738

Rickels MR, Norris AW, Hull RL (2020) A tale of two pancreases: exocrine pathology and endocrine dysfunction. Diabetologia 63(10):2030–2039
Sarkany RPE, Moreland BH (1985) Enhancement of the autocatalytic activation of trypsinogen to trypsin by bile and bile acids. Biochim Biophys Acta Gen Subj 839(3):262–267
Schaffer AE, Freude KK, Nelson SB, Sander M (2010) Nkx6 transcription factors and Ptf1a function as antagonistic lineage determinants in multipotent pancreatic progenitors. Dev Cell 18(6):1022–1029
Seymour PA, Collin CA, Egeskov-Madsen A l R, Jørgensen MC, Shimojo H, Imayoshi I et al (2020) Jag1 modulates an oscillatory Dll1-Notch-Hes1 signaling module to coordinate growth and fate of pancreatic progenitors. Dev Cell 52(6):731–747.e8
Shih HP, Seymour PA, Patel NA, Xie R, Wang A, Liu PP et al (2015) A gene regulatory network cooperatively controlled by Pdx1 and Sox9 governs lineage allocation of foregut progenitor cells. Cell Rep 13(2):326–336
Shih HP, Panlasigui D, Cirulli V, Sander M (2016) ECM signaling regulates collective cellular dynamics to control pancreas branching morphogenesis. Cell Rep 14(2):169–179
Soave D, Miller MR, Keenan K, Li W, Gong J, Ip W et al (2014) Evidence for a causal relationship between early exocrine pancreatic disease and cystic fibrosis–related diabetes: a Mendelian randomization study. Diabetes 63(6):2114–2119
Solar M, Cardalda C, Houbracken I, Martín M, Maestro MA, De Medts N et al (2009) Pancreatic exocrine duct cells give rise to insulin-producing β cells during embryogenesis but not after birth. Dev Cell 17(6):849–860
Sznurkowska MK, Hannezo E, Azzarelli R, Rulands S, Nestorowa S, Hindley CJ et al (2018) Defining lineage potential and fate behavior of precursors during pancreas development. Dev Cell 46(3):360–375.e5
Tucker JA, Spock A, Spicer SS, Shelburne JD, Bradford W (2003) Inspissation of pancreatic zymogen material in cystic fibrosis. Ultrastruct Pathol 27(5):323–335
van Roy F, Berx G (2008) The cell-cell adhesion molecule E-cadherin. Cell Mol Life Sci 65(23):3756–3788
Villasenor A, Chong DC, Henkemeyer M, Cleaver O (2010) Epithelial dynamics of pancreatic branching morphogenesis. Development 137(24):4295–4305
Watt AJ, Zhao R, Li J, Duncan SA (2007) Development of the mammalian liver and ventral pancreas is dependent on GATA4. BMC Dev Biol 7(1):37
Wilschanski M, Novak I (2013) The cystic fibrosis of exocrine pancreas. Cold Spring Harb Perspect Med 3(5):a009746
Xuan S, Sussel L (2016) GATA4 and GATA6 regulate pancreatic endoderm identity through inhibition of hedgehog signaling. Development 143(5):780–786
Yue B (2014) Biology of the extracellular matrix: an overview. J Glaucoma 23:S20–S23
Zhang Q, Davenport JR, Croyle MJ, Haycraft CJ, Yoder BK (2005) Disruption of IFT results in both exocrine and endocrine abnormalities in the pancreas of Tg737orpk mutant mice. Lab Investig 85(1):45–64
Zhang H, Ables ET, Pope CF, Washington MK, Hipkens S, Means AL et al (2009) Multiple, temporal-specific roles for HNF6 in pancreatic endocrine and ductal differentiation. Mech Dev 126(11):958–973
Zhou Q, Law AC, Rajagopal J, Anderson WJ, Gray PA, Melton DA (2007) A multipotent progenitor domain guides pancreatic organogenesis. Dev Cell 13(1):103–114

β-Cell Heterogeneity and Plasticity

Hyo Jeong Yong and Yue J. Wang

Abstract The existence of functionally diverse and plastic β cells in islets of Langerhans has been reported since the 1980s. Recently, high-resolution technologies have advanced our understanding of β-cell heterogeneity and plasticity. Here, we define plasticity broadly as dynamic changes in cellular phenotypes and heterogeneity as differences in cellular behaviors. Individual β cells react differently to environmental challenges and act together to maintain β-cell mass and glucose homeostasis within a narrow range of 70–140 mg/dL. During the progress of diabetes, this elaborate balance is disrupted, and a lack of β-cell compensation leads to dysregulated blood glucose. In this chapter, we assess β-cell stress that instigates increased β-cell heterogeneity and adaptive β-cell responses such as proliferation, dedifferentiation, maturity, and insulin secretion. We also discuss the maturity, electrical activity, and insulin secretion of well-characterized β-cell subgroups. Finally, we touch upon the plasticity of other non-β pancreatic cells and their cooperation with β cells to maintain homeostasis.

Keywords Pancreas · Islet of Langerhans · β cells · Heterogeneity · Plasticity · Diabetes

Abbreviations

ARX	Aristaless-related homeobox
ATAC-seq	Assay for transposase-accessible chromatin with sequencing
BIRC5	Baculoviral IAP repeat containing 5
BrdU	Bromodeoxyuridine

H. J. Yong · Y. J. Wang (✉)
Department of Biomedical Sciences, College of Medicine, Florida State University, Tallahassee, FL, USA
e-mail: hyojeong.yong@med.fsu.edu; julia.wang@med.fsu.edu

M. L. Golson (ed.), *Molecular and Cell Biology of Pancreas Development, Function and Regeneration*, Advances in Anatomy, Embryology and Cell Biology 239, https://doi.org/10.1007/978-3-031-62232-8_3

CFTR	Cystic fibrosis transmembrane conductance regulator
CK	Cytokeratin
DNMT1	DNA methyltransferase 1
ER	Endoplasmic reticulum
ERO1LB	Endoplasmic reticulum oxidoreductase 1-like protein β
eIF2α	Eukaryotic initiation factor-2α
FTH1	Ferritin heavy chain 1
FTH1P3	Ferritin heavy chain 1 pseudogene 3
FTL	Ferritin light chain
FAD	Flavin adenine dinucleotide
FOXM1	Forkhead box M1
FOXO1	Forkhead box O1
GWAS	Genome-wide association studies
G6PC2	Glucose-6-phosphatase catalytic subunit 2
GLUT2	Glucose transporter 2
GSIS	Glucose-stimulated insulin secretion
HNF1A	Hepatocyte nuclear factor 1 homeobox A
IRE1	Inositol-requiring enzyme 1
INS	Insulin
XRM	Integrated X-ray microscopy
IAPP	Islet amyloid polypeptide
Islet	Islet of Langerhans
MAFA	MAF BZIP Transcription Factor A
MAPK	Mitogen-activated protein kinase
GM2	Monosialodihexosylganglioside
MIP-GFP	Mouse insulin I promoter-driven GFP
MUC1	Mucin 1
NEUROG3	Neurogenin 3
NEUROD1	Neuronal differentiation 1
NAD(P)H	Nicotinamide adenine dinucleotide phosphate
NKX2.2	NK2 homeobox 2
OCT4	Octamer-binding transcription factor 4
PAX4	Paired box 4
PTF1A	Pancreas transcription factor 1 subunit α
PDX1	Pancreatic and duodenal homeobox 1
PP	Pancreatic polypeptide
PERK	PKR-like endoplasmic reticulum kinase
PCNA	Proliferating cell nuclear antigen
PKR	Protein kinase R
REG3A	Regenerating family member 3 α
RBP4	Retinol binding protein 4
SEM	Scanning electron microscope
SCTR	Secretin receptor
MIMS-EM	SEM with multi-isotope mass spectrometry
SQSTM1	Sequestosome 1

PRSS1	Serine protease 1
STAT3	Signal transducer and activator of transcription 3
smFISH	Single-molecule fluorescence in situ hybridization
SLC4A4	Solute carrier family 4 member 4
SOX9	SRY-box transcription factor 9
ST8SIA1	ST8 α-N-acetyl-neuraminide α-2,8-sialyltransferase 1
SI	Stimulation index
TFF1	Trefoil factor 1
T1D	Type 1 diabetes
T2D	Type 2 diabetes
UPR	Unfolded protein response
UCN3	Urocortin 3
XBP1	X-box binding protein 1
ZF	Zucker fatty
$[Ca^{2+}]$	Cytosolic Ca^{2+} concentration

1 Introduction

The pancreas is composed of exocrine and endocrine components. The exocrine tissue contains acinar and ductal cells. Together, exocrine cells release and deliver digestive enzymes to the gut. Islets of Langerhans (islets) are clusters in which endocrine cells are gathered. Five major types of endocrine cells (α, β, δ, pancreatic polypeptide (PP), and ε) comprise islets. Together, the islet endocrine cells regulate blood glucose by secreting hormones into the circulatory system. The pancreas also contains other stromal cell types, including endothelial cells, fibroblasts, neurons, and immune cells. Among these pancreatic cells, β cells play the most pivotal role in maintaining glucose homeostasis by producing insulin.

Intracellular lipofuscin body accumulation implies that human β cells are long-lived cells with limited proliferative capacity (Cnop et al. 2010). Proliferating β cells are indeed observed primarily in subjects younger than 30 years old (Perl et al. 2010). Similarly, the lifespan and the turnover rates of mouse β cells were assessed by integrated X-ray microscopy (XRM) and scanning electron microscope (SEM) with multi-isotope mass spectrometry (MIMS-EM) (Arrojo et al. 2019). The authors found that around 60% of mouse β cells are born by postnatal day 45 and remain stable through the end of the 26-month chasing period (Arrojo et al. 2019; Arrojo 2021).

Although they live extraordinarily long lives, β cells have fluxes in mass and function to adapt to metabolic demands and environmental changes (Kulkarni et al. 2012; Baeyens et al. 2018). Even during periods of homeostasis, cellular and spatial heterogeneity of β cells in glucose sensing and insulin secretion occurs (Salomon and Meda 1986; Stefan et al. 1987; Heimberg et al. 1993). In mice, virgin β cells are an immature β-cell subgroup that expresses insulin but not markers of maturity such as *Urocortin 3* (*Ucn3*), *glucose-6-phosphatase catalytic subunit 2* (*G6pc2*), *endoplasmic reticulum oxidoreductase 1-like protein β* (*Ero1lb*), and *glucose transporter 2* (*Glut2*). This subgroup was discovered in a neogenic niche at the periphery of islets (van der Meulen et al. 2017; Lee et al. 2021) (See "Virgin β cells" below for more information). In zebrafish, younger, less mature β cells occupy anterior regions in islets and display higher proliferation rates and diminished insulin response to elevated glucose. In contrast, older, more mature β cells reside in posterior areas of islets and are more responsive to glucose stimulation (Singh et al. 2017). With the advance in high-resolution technologies, recent studies have further dissected the molecular determinants of β-cell heterogeneity. Applying single-cell RNA-sequencing technology to human islets, several research groups have dissected subpopulations of β cells and deciphered their roles in glucose homeostasis (Baron et al. 2016; Li et al. 2016; Wang et al. 2016a; Muraro et al. 2016; Segerstolpe et al. 2016; Xin et al. 2016; Enge et al. 2017; Lawlor et al. 2017; Avrahami et al. 2020; Yong et al. 2021, 2022).

Here we provide an overview of the heterogeneity and plasticity of β cells and discuss how these features improve our understanding of their roles in maintaining functional β-cell mass for glucose regulation. Firstly, we address the impact of β-cell stress, including aging and endoplasmic reticulum (ER) stress, on β-cell heterogeneity and plasticity. Secondly, we highlight β-cell heterogeneity and plasticity in proliferation, dedifferentiation, maturity, and insulin secretion. Thirdly, we discuss newly characterized β-cell subgroups and their functions. Lastly, transdifferentiation from other pancreatic cells such as α, acinar, and ductal cells to β cells is described. Gaining a better understanding of β-cell heterogeneity and plasticity will shed light on how different β-cell populations (mal)adapt to environmental challenges and how these processes contribute to β-cell regeneration and disease progression.

2 β-Cell Stress and β-Cell Heterogeneity/Plasticity

Increased metabolic demands induced by diabetes, obesity, injury, and pregnancy elicit β-cell heterogeneity and plasticity. To adapt to these adverse environments and to maintain glucose homeostasis, β cells react by activating short-term mechanisms to minimize acute damage and long-term responses to gain resistance to chronically harsh conditions. Here, we highlight aging as one of the universal processes associated with increased β-cell heterogeneity and plasticity. We further describe the ER stress response that acts as a nexus in the convergent pathway that controls the β-cell adaptation to different environments (Fig. 1).

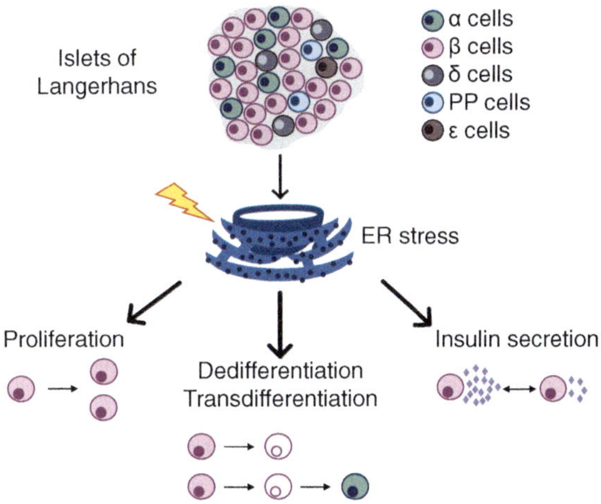

Fig. 1 Manifestations of β-cell heterogeneity and plasticity. β cells may proliferate, dedifferentiate/transdifferentiate, and alter their insulin secretory capacity to adapt to environmental challenges. These environmental challenges are mediated through ER stress. Green, pink, gray, blue, and brown cells represent α, β, δ, PP, and ε cells, respectively. Light pink cells with light nuclei represent progenitor cells

3 Aging

β cells undergo both maturation and senescence with aging, displaying more prominent heterogeneity with time (Aguayo-Mazzucato et al. 2017). Aging is a risk factor for type 2 diabetes (T2D), potentially due in part to the reduced β-cell proliferation and function in older individuals (Perl et al. 2010; Chang and Halter 2003; Meier et al. 2008; De Tata 2014).

β-cell proliferation declines with age. Human β-cell proliferation occurs mainly in the first few years of life. At any given time, about 3% of β cells are proliferating in children less than 2 years of age, and the replication rate declines to ~0.1% in older adults (Perl et al. 2010; Meier et al. 2008, 2010; Kassem et al. 2000; Wang et al. 2016b). A single-cell mass cytometry study of human islets demonstrated that a β-cell cluster with high Ki67 expression negatively correlated with donor age (Wang et al. 2016b). Furthermore, β-cell regeneration in response to pancreatic injury is negatively correlated with age. Young mice partially restored β-cell mass after pancreatic damage by stimulating regenerative signals across the entire pancreas, but by comparison, aged mice showed a remarkably reduced capacity for regeneration after injury (Dor et al. 2004; Nir et al. 2007; Cano et al. 2008; Rankin and Kushner 2009; Thorel et al. 2010; Afelik and Rovira 2017).

β-cell function also declines with age. Humans display a decreased insulin secretion capacity with age (Chen et al. 1985; Ihm et al. 2006). Like humans, older rodents secrete diminished insulin in response to high glucose (Ihm et al. 2007). Additionally, the transcriptional levels of genes related directly to β-cell function,

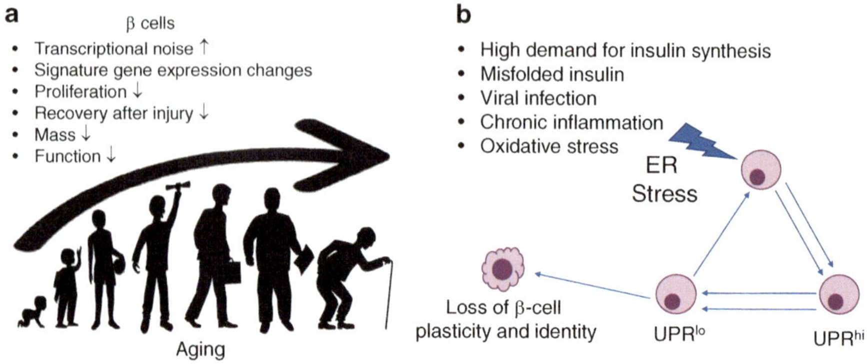

Fig. 2 Examples of common biological processes associated with β-cell heterogeneity and plasticity. (**a**) With age, an increase in transcriptional noise, changes in signature gene expression, a decrease in proliferation, a reduction in recovery after pancreatic injury, and diminished mass and function occur in β cells. (**b**) β cells recover from ER stress by activating high and low levels of unfolded protein response (UPR) (UPR^{hi} and UPR^{lo}) genes. However, β cells lose their identity and the ability to expand under persistent low levels of UPR activity

such as *insulin 2* (*Ins2*), *Glut2, and pancreatic and duodenal homeobox 1* (*Pdx1*), were lower in older rodents than in young rodents (Ihm et al. 2007).

Single-cell transcriptomics on islets from multiple donors reveals that gene signatures of α and β cells from children did not match those from adults (Wang et al. 2016a). In young donors, β cells expressed α-cell signature genes, and α cells expressed mature β-cell markers (Wang et al. 2016a). Aged β cells display an increase in oxidative stress-related mutations and elevated transcriptional noise. These molecular changes may contribute to a loss of β-cell identity and age-related degeneration (Fig. 2a) (Enge et al. 2017; Chang and Halter 2003; De Tata 2014; Kim et al. 2013; Solas et al. 2013). Single-cell RNA sequencing from cynomolgus monkeys revealed that aged β cells differentially express four times more aging-associated genes than aged α cells (Li et al. 2021). The β-cell genes whose expression decreased with age were related to peptide biosynthesis and secretory granule biogenesis, while genes whose expression increased with age were related to ER stress and the unfolded protein response (UPR) (Li et al. 2021).

4 ER Stress

β cells, due to their roles in continued insulin secretion in response to glucose perturbation, are consistently under high demand for insulin synthesis and are prone to accumulate unfolded or misfolded insulin. These adverse conditions lead to ER stress, which plays a central role in regulating the plasticity and heterogeneity in β-cell function and maturity (Hwang and Qi 2018; Sahin et al. 2021; Sharma et al. 2021). In type 1 diabetes (T1D), β-cell dysfunction is associated with excessive ER stress (Hwang and Qi 2018; Sahin et al. 2021; Sharma et al. 2021). Similarly, in T2D, impaired β-cell function may be driven by higher ER stress (Marchetti et al. 2007; Engin et al. 2014). UPR-related variants identified by genome-wide association studies (GWAS) are also strongly correlated with increased β-cell dysfunction and T2D risk (Chu et al. 2007; Cho et al. 2011; Liu et al. 2016; Wood et al. 2017).

UPR is activated to mitigate ER stress, and basal levels of UPR are high in β cells (Sahin et al. 2021; Schroder and Kaufman 2005; Bernales et al. 2006). Inositol-requiring enzyme 1 (IRE1), protein kinase R (PKR)-like endoplasmic reticulum kinase (PERK), and activating transcription factor 6 (ATF6) are ER transmembrane proteins involved in UPR (Hetz 2012). Deleting *Ire1a* in NOD mice induced dedifferentiation of β cells, enabling their survival against autoimmunity. This survival may come from decreased autoantigens and MHC class I components through dedifferentiation (Lee et al. 2020). However, introducing an inactivating mutation in a downstream target of PERK signaling, eukaryotic initiation factor-2α (eIF2α), reduced the β-cell mRNA expression of identity markers such as *Pdx1, MAF BZIP Transcription Factor A* (*MafA*), *Ins1, Ins2, islet amyloid polypeptide* (*Iapp*), *and Glut2* (Back et al. 2009). The deletion of *X-box binding protein 1* (*Xbp1*), an ER stress response factor activated by IRE1α, led to β-cell to α-cell transdifferentiation in mice (Lee et al. 2022). These results suggest that UPR is a double-edged sword that can be both beneficial and detrimental to β-cell homeostasis, depending on the context.

Recently, multiple single-cell RNA-sequencing studies have demonstrated the existence of β-cell subpopulations with higher ER stress (Baron et al. 2016; Muraro et al. 2016; Wang and Kaestner 2019). A single-cell RNA-sequencing analysis identified two clusters of human β cells based on differential gene expression demonstrating changes in β-cell function and ER stress pathways (Baron et al. 2016). One of these populations highly expressed β-cell function-related genes, while the other highly expressed ER-stress-inducible genes. Intriguingly, the other human islet endocrine cell types did not show this heterogeneity of functional and ER stress gene expression (Baron et al. 2016). Another single-cell transcriptomics study reported that the most significant genes affecting β-cell heterogeneity are *sequestosome 1* (*SQSTM1*) *and three ferritin subunits: ferritin heavy chain 1 pseudogene 3* (*FTH1P3*), *ferritin heavy chain 1* (*FTH1*), *and ferritin light chain* (*FTL*) (Muraro et al. 2016). These genes are all involved in ER stress (Orino et al. 2001; Rantanen et al. 2013; Zhou et al. 2015).

A pseudotime ordering of human β cells indicates that they cycle through different states based on their endocrine function and UPR (Xin et al. 2018) (Fig. 2b). In one state, β cells express extremely high insulin levels and concomitantly high ER stress. Thereafter, the β cells enter a recovery state marked by activated UPR, decreased insulin biosynthesis, and increased proliferation (Sharma et al. 2015). As UPR gradually reduced in the recovering β cells, they started to prepare for insulin biosynthesis. After recovery, the β cells resume insulin secretion (Xin et al. 2018). In agreement with the pseudotime analysis, β cells under ER stress display transcriptional and translational alterations. When MIN6 cells underwent ER stress, β-cell marker gene expression decreased in direct relation to the duration of ER stress. These β cells adapted to prolonged ER stress by significantly upregulating UPR genes. When chronic ER stress resolved, MIN6 cells recovered β-cell identity genes. However, with repetitive ER stress cycles, MIN6 cells lost their plasticity and β-cell identity (Fig. 2b). Remarkably, single-cell RNA sequencing of β cells from T1D patients displayed similar gene expression changes as β cells under ER stress (Chen et al. 2022). These results may suggest that the attempts of a few remaining β cells to compensate for the loss of a majority of functional β-cell mass.

5 The Major Manifestations of β-Cell Heterogeneity/ Plasticity

Three major adaptations that human β cells undergo throughout their lifetime in response to environmental changes are proliferation, dedifferentiation or transdifferentiation, and changes in insulin secretion. The first adaptation is proliferation which is the replication of existing β cells and occurs primarily in the first few years of life (Meier et al. 2008, 2010; Kassem et al. 2000). The second adaption is dedifferentiation or transdifferentiation. Dedifferentiation is the conversion of mature β cells to progenitor-state cells through the loss of β-cell-specific marker gene expression, which can be triggered by cellular stress (Wang and Zhang 2021). Transdifferentiation is the process whereby β cells directly convert to another mature cell type without reverting to progenitor states and is also related to cellular stress (Piran et al. 2014; Gutiérrez et al. 2017).

The third adaptation β cells implement is variations in insulin-release capacities (Kiekens et al. 1992; Van Schravendijk et al. 1992). For example, responsive β cells react to elevated glucose by increasing insulin secretion. In contrast, unresponsive β cells are insensitive to changes in glucose concentrations (Giordano et al. 1991). Depending on conditions, approximately 25% of β cells switch between a responsive and nonresponsive cell state (Giordano et al. 1991). Recently, a study using patch-seq directly linked differential gene expression patterns and heterogeneous electrophysiological functions of human β cells at the single-cell level (Camunas-Soler et al. 2020). This group identified heterogeneous β-cell function in Ca^{2+} currents, Na^+ currents, and exocytosis (Camunas-Soler et al. 2020). In particular, they revealed that Na^+ currents and exocytosis are negatively correlated with the expression of retinol binding protein 4 (RBP4), a known heterogeneously expressed gene in β cells (Baron et al. 2016; Segerstolpe et al. 2016; Camunas-Soler et al. 2020). In the following paragraphs, we expound on β-cell heterogeneity and plasticity as manifested by proliferation, dedifferentiation or transdifferentiation, and insulin secretion.

6 Proliferation

β-cell proliferation increases in response to T1D, T2D, pregnancy, obesity, and injury (Table 1). During T1D progression, β-cell mass progressively but non-linearly decreases. T1D progression can be divided into four steps, based on β-cell mass and function: prediabetes, onset, remission, and overt T1D (Chen et al. 2017). Although β-cell death gradually increases preceding a diabetes diagnosis, β-cell mass is unchanged or slightly decreases during the prediabetic phase in T1D (Gianani et al.

Table 1 Changes in β-cell mass and proliferation in rodents and humans

		Diabetes	Pregnancy	Obesity	Injury
Mass	Rodents	From 70% to 90% reduction (Sreenan et al. 1999; Sherry et al. 2006)	From 2- to 3.8-fold increase (Karnik et al. 2007; Rieck et al. 2009)	Twofold increase (Pick et al. 1998; Hull et al. 2005)	From 17% recovered to 12% increase (Nir et al. 2007; Rankin and Kushner 2009; Thorel et al. 2010)
	Humans	Up to 90% reduction (Gepts 1965; Kloppel et al. 1984; Butler et al. 2007)	From 1.4- to 2-fold increase (Van Assche et al. 1978; Butler et al. 2010)	From 1.5- to 2-fold increase (Butler et al. 2003; Hanley et al. 2010; Saisho et al. 2013)	Not recovered (Menge et al. 2008)
Proliferation	Rodents	Tenfold increase (Sreenan et al. 1999; Sherry et al. 2006)	From 2- to 10-fold increase (Green et al. 1981 Parsons et al. 1992; Rieck et al. 2009)	1.4-fold increase (Pick et al. 1998; Hull et al. 2005)	From 2.5- to 10-fold increase (Dor et al. 2004; Nir et al. 2007; Rankin and Kushner 2009)
	Humans	No difference (Butler et al. 2007) From 10- to 100-fold increase (Meier et al. 2006; Willcox et al. 2010)	No difference (Butler et al. 2010)	No difference (Butler et al. 2003; Saisho et al. 2013) 1.4-fold increase (Hanley et al. 2010)	No difference (Menge et al. 2008)

2006; In't Veld et al. 2007; Herold et al. 2015; Rodriguez-Calvo et al. 2017). Some studies have even reported increased β-cell proliferation upon inflammation in prediabetic mice or a twofold elevated β-cell density in nondiabetic autoantibody-positive human donors with insulitis (Sreenan et al. 1999; Sherry et al. 2006; Campbell-Thompson et al. 2016). Since human pancreata used for research are almost always from deceased organ donors, results from human pancreata are complicated by the inability to perform longitudinal studies. Human studies on β-cell density are further complicated by a wide variability of β-cell density, even in control subjects. However, these results may explain a delay in decreased β-cell mass during the prediabetic phase of T1D in some individuals. At the onset of T1D, β-cell mass can be reduced by more than 90% (Gepts 1965; Junker et al. 1977; Gepts and De Mey 1978; Kloppel et al. 1984; Butler et al. 2007). Compared to β cells from nondiabetic subjects, recently diagnosed donors with T1D displayed elevated β-cell proliferation, but this increase was insufficient to compensate for the massive β-cell culling by the immune system (Meier et al. 2006; Willcox et al. 2010). During T1D remission (also called the honeymoon phase), hyperglycemia transiently returns to normoglycemia due to functional compensation by the remaining β cells or by insulin treatment. While a slightly increased β-cell mass and proliferation were observed during T1D remission as compared to the onset of T1D, significant β-cell regeneration has not been detected (Sherry et al. 2006; Willcox et al. 2010; Chmelova et al. 2015; Fonolleda et al. 2017; Tang et al. 2019). During the overt diabetes stage, the relationship between β-cell mass and T1D duration is ambiguous (Butler et al. 2007; Lam et al. 2017), and no correlation between β-cell proliferation and diabetes duration has been observed (Lam et al. 2017; Meier et al. 2005; Oram et al. 2019).

Multiple aspects of obesity contribute to T2D onset and progression. Increased peripheral fat mass elicits insulin resistance, resulting in β-cell dysfunction due to high metabolic demands in susceptible populations (Anderson et al. 2003; Kasuga 2006; Kim and Park 2018). Obesity is also accompanied by the accumulation of pancreatic fat, which may contribute to β-cell dysfunction (Lee et al. 1994, 2010). To compensate for insulin resistance, β-cell mass increases in obese individuals. A 3.7-fold increase in β-cell mass was detected in young obese mice, and increased β-cell proliferation was observed in nondiabetic Zucker fatty (ZF) rats (Pick et al. 1998; Hull et al. 2005; Tschen et al. 2009). In humans, β-cell mass increased twofold in nondiabetic subjects with obesity (Brandhorst et al. 1995; Butler et al. 2003; Hanley et al. 2010; Saisho et al. 2013). However, the primary mechanisms contributing to β-cell mass elevations during obesity need further study. In both rodents and humans, the mechanisms by which β-cell mass increases in response to obesity are unclear, but both hypertrophy and proliferation appear to play a role in β-cell mass expansion (Brandhorst et al. 1995; Butler et al. 2003; Saisho et al. 2013; Matsumoto et al. 2004; Linnemann et al. 2014).

Proliferation is also an underlying mechanism for mouse β-cell regeneration after pancreas injury. Single-cell RNA sequencing of partial pancreatectomy mouse models identified proliferative β-cell clusters (Tatsuoka et al. 2020). This population was in S and G2/M phases and exhibited high expression levels of *proliferating cell nuclear antigen* (*Pcna*) and *Mki67*. While genes related to β-cell function such as *Ins1*, *Ins2*, and *Ucn3* were downregulated in proliferating β cells, immaturity and progenitor markers such as *Octamer-binding transcription factor 4* (*Oct4*) or *Neurogenin 3* (*Neurog3*), as well as other endocrine cell markers, were not detected (Tatsuoka et al. 2020). In addition, mice lacking the proliferation-associated transcription factor Forkhead box M1 (Foxm1) in the pancreas have reduced β-cell proliferation and β-cell mass after recovering from a 60% partial pancreatectomy (Ackermann Misfeldt et al. 2008). In humans, however, partial pancreatectomy patients did not exhibit increased β-cell proliferation, even after up to 50% removal of pancreas mass (Menge et al. 2008). However, the mean age of patients in that study was 51.9, and partial pancreatectomy does not elicit β-cell proliferation in mice older than 12 months (Rankin and Kushner 2009). Furthermore, β-cell proliferation was examined, on average, 1.2 years after the initial surgery, and the window of increased cell division may have been missed (Menge et al. 2008).

During pregnancy, maternal metabolism is altered due to the energy demand of the fetus. To mitigate metabolic changes, prolactin and placental lactogen counterbalance the insulin resistance that develops during pregnancy and prevent maternal hyperglycemia by driving β-cell expansion (Van Assche et al. 1978; Parsons et al. 1992; Huang et al. 2009; Rieck and Kaestner 2010). In animal models, the overexpression of prolactin and placental lactogen during pregnancy stimulates a three to fivefold β-cell mass increase. Upregulation or overexpression of prolactin and lactogen is necessary and sufficient for stimulating β-cell proliferation (Parsons et al. 1992, 1995; Huang et al. 2009; Sorenson and Brelje 1997; Vasavada et al. 2000; Freemark et al. 2002; Karnik et al. 2007; Rieck et al. 2009; Kim et al. 2010; Zhang et al. 2010; Ernst et al. 2011). Gene expression studies during pregnancy have revealed that pregnancy-induced β-cell proliferation appears to act through the FOXM1-baculoviral IAP repeat containing 5 (Birc5) pathway (Rieck et al. 2009; Layden et al. 2010). Strikingly, increased β-cell mass returns to pre-pregnancy levels within 10 days after birth through decreased β-cell proliferation and increased apoptosis (Van Assche et al. 1978; Scaglia et al. 1995; Butler et al. 2010).

Historically, β-cell proliferation studies have relied on rodent models. Even though we have gained valuable knowledge from rodent β cells, human and rodent β cells differ in several essential characteristics. For example, human β-cell mass increases only moderately in response to pregnancy (from 1.4- to 2-fold) and obesity (from 1.5- to 2-fold), as compared to the higher β-cell mass increase in rodents during similar metabolic processes (pregnancy from 2- to 3.8-fold; obesity from 1.3- to 3.7-fold) (Pick et al. 1998; Hull et al. 2005; Butler et al. 2003, 2010; Hanley et al. 2010; Saisho et al. 2013; Van Assche et al. 1978; Rieck and Kaestner 2010; Karnik et al. 2007; Casimiro et al. 2021; Mosser et al. 2015; Golson et al. 2010). β-cell mass recovery after an injury has been observed in rodents, but significant β-cell regeneration has not been detected in humans (Dor et al. 2004; Nir et al. 2007; Xu et al. 2008; Zhou et al. 2008; Criscimanna et al. 2011; Pan et al. 2013). Therefore, more evidence is needed to fill gaps in the knowledge regarding the human β-cell proliferative response. Table 1 summarizes the changes in beta cell mass and proliferation observed in rodents and humans during different (patho)physiological processes.

7 Dedifferentiation/Transdifferentiation

In T2D, β-cell loss ranges from 24% to 65%, and islet size is reduced (Butler et al. 2003; Hanley et al. 2010; Kloppel et al. 1985; Clark et al. 1988; Sakuraba et al. 2002; Yoon et al. 2003; Deng et al. 2004; Rahier et al. 2008; Inaishi et al. 2016). Several studies argue that the major mechanism of β-cell loss in T2D is increased apoptosis (Marchetti et al. 2007, 2004; Butler et al. 2003; Hanley et al. 2010; Yoneda et al. 2013). However, other researchers propose that β cells undergo dedifferentiation rather than apoptosis during T2D pathogenesis (Lin et al. 2011; Talchai et al. 2012; Guo et al. 2013; Marselli et al. 2014; Spijker et al. 2015; Cinti et al. 2016; Lu et al. 2018; Sun et al. 2019). In *Lepr*$^{db/db}$ mice, reduced expression of the key β-cell transcription factors MAFA and NKX6.1 was detected in β cells during hyperglycemia development (Guo et al. 2013). In mice lacking insulin receptors (GIRKO mice) with T2D, β cells lost insulin and PDX1 expression, and ectopic α cells were present in islets (Lin et al. 2011). Similarly, in a model of hyperglycemia caused by *Forkhead box O1* (*Foxo1*) ablation, β cells dedifferentiated into progenitor-like cells, and a subset of the progenitor-like cells gave rise to α cells (Talchai et al. 2012). The same group confirmed β-cell dedifferentiation in *Lepr*$^{db/db}$ mice (Talchai et al. 2012).

In humans with T2D, loss of β-cell identity is characterized by degranulation, a decrease in PDX1, MAFA, and NKX6.1 expression, and a reduction in β-cell number (Guo et al. 2013; Marselli et al. 2014). In addition, dedifferentiated endocrine cells are increased by threefold in T2D subjects compared with nondiabetic subjects, and dedifferentiation from β cells to progenitor-like cells was observed in T2D (Spijker et al. 2015; Cinti et al. 2016; Sun et al. 2019). These progenitor-like cells are partially converted to α- or δ-like cells (Spijker et al. 2015; Cinti et al. 2016). Recent evidence suggests that polycomb silencing regulates β-cell dedifferentiation (Lu et al. 2018). Loss of β-cell-specific polycomb (Eed/PRC2), which maintains β-cell identity by repressing endocrine progenitor genes, leads to β-cell dedifferentiation followed by T2D (Lu et al. 2018). In addition to dedifferentiation, transdifferentiation has also been reported as a mechanism contributing to β-cell fate alterations under metabolic stress (Piran et al. 2014; Spijker et al. 2015; Cinti et al. 2016; Brereton et al. 2014). However, the distinction between dedifferentiation and transdifferentiation might be a semantic division rather than distinct biological processes (Mills et al. 2019).

8 Maturity and Insulin Secretion

Not all β cells are mature, respond similarly to glucose stimulation, or have homogeneous metabolic activity. β-cell function also fluctuates during diabetes progression. Using a lentivirus dual fluorescent reporter for PDX1/INS1, the kinetics of β-cell maturation were uncovered in both human and mouse β cells (Szabat et al.

2011). In human islets, immature PDX1^{+}/INS^{low} and mature PDX1^{high}/INS^{high} were identified. The endocrine function-related genes, such as glucose transporters and solute carrier families, were enriched in mature PDX1^{high}/INS^{high} β cells (Szabat et al. 2011). PDX1^{+}/INS^{low} mouse islets expressed high levels of glucagon and genes associated with β-cell development (Szabat et al. 2011). Single-cell transcriptomics of human islets revealed heterogeneous insulin expression in human β cells (Li et al. 2016). Likewise, individual rodent β cells exhibited different expression levels of insulin mRNA and protein (Van Schravendijk et al. 1992; Bengtsson et al. 2005; Scarl et al. 2019). Insulin secretion can vary considerably between β cells (Wojtusciszyn et al. 2008). Transgenic mice expressing mouse insulin I promoter-driven GFP (MIP-GFP) revealed three populations of β cells: GFP^{low}, GFP^{medium}, and GFP^{high}. GFP^{medium} and GFP^{high} β cells comprised 90% of this population and had more insulin granules and higher levels of insulin secretion. In contrast, the GFP^{low} population made up only 10% of β cells and demonstrated fewer insulin granules, lower levels of insulin secretion, and decreased glucose-stimulate insulin secretion (GSIS) (Aguayo-Mazzucato et al. 2017; Katsuta et al. 2012). The proportion of cells secreting insulin varies in response to changing glucose levels. At 2 mM glucose, 20% of adult human β cells secrete 90% of released insulin. At 20 mM glucose, 5% of β cells release more insulin than average secreted insulin per cell, whereas 10% of β cells rarely release insulin at 20 mM glucose (Wojtusciszyn et al. 2008). β cells also display heterogeneous metabolic activity, measured using an autofluorescence signal that correlates with flavin adenine dinucleotide (FAD) and nicotinamide adenine dinucleotide phosphate (NAD(P)H) levels (Kiekens et al. 1992; Scarl et al. 2019; Piston et al. 1999). β cells with high NAD(P)H autofluorescence contained more insulin granules. Interestingly, β cells with fewer insulin granules were resistant to cell death in the face of immune-cell attack (Scarl et al. 2019; Van De Winkel and Pipeleers 1983; Hellman et al. 1992; Rui et al. 2017).

β cells also display plasticity in insulin secretion functions during diabetes progression. In the T1D remission period, β cells produce more C-peptide and proinsulin and show partial recovery, such as re-granulation, compared to β cells in the onset of T1D and overt T1D (Chmelova et al. 2015; Jackson et al. 1940; Brush 1944; Block et al. 1973; Karges et al. 2004; Abdul-Rasoul et al. 2006; Scholin et al. 2011; Kaas et al. 2012). Obese nondiabetic subjects exhibit higher fasting plasma insulin levels and a greater insulin response to glucose stimulus compared to lean nondiabetic subjects (Brandhorst et al. 1995; Perley and Kipnis 1966; Seltzer et al. 1967; Polonsky et al. 1988a; Lillioja et al. 1993; Ferrannini et al. 1997; Camastra et al. 2005; Mezghenna et al. 2011). After the onset of T2D, obvious β-cell dysfunction is evident. Insulin secretion capacity decreased up to 97%, and the dynamics of insulin release are altered due to disrupted first-phase response and oscillations of cytosolic Ca^{2+} concentration ($[Ca^{2+}]$) (Brunzell et al. 1976; Lang et al. 1981; Ward et al. 1984; Polonsky et al. 1988b; Osei et al. 1997; Holman 1998; Jensen et al. 2002; Gastaldelli et al. 2004; Ferrannini et al. 2005). Interestingly, β cells isolated from T2D mice recovered normal insulin production capacity when incubated in normoglycemic media (Alarcon et al. 2016).

Recently, β-cell functional heterogeneity between the sexes was reported. When female and male islet function and molecular programs were compared, female islets had a higher insulin secretion capacity than male islets (Yong et al. 2022). Single-cell RNA-sequencing analysis revealed that female endocrine cells have higher transcriptional, translational, and metabolic pathway activities, while male endocrine cells have higher cell stress signatures (Yong et al. 2022). The distinct molecular signatures of female and male endocrine cells may explain why female islets have a higher secretion capacity than male islets. The sex-divergent islet function may explain the higher risks of diabetes in males compared with females (Yong et al. 2022).

Finally, heterogeneity in β-cell function due to circadian clock regulation and the impact of an impaired circadian system on T2D have been explored (Seshadri and Doucette 2021). β-cell insulin secretion transcriptional profiles vary, depending on whether the circadian clock is in an active (awake) or inactive (asleep) phase (Perelis et al. 2015; Rakshit et al. 2016).

9 Functional β-Cell Subgroups

In the preceding section, we delineated the existence of β-cell heterogeneity and plasticity. In this section, we will focus on well-characterized β-cell subgroups, their identifying markers, and their functional differences in maturity, electrical activity, and insulin secretion (Fig. 3). We will further discuss the contribution of β-cell heterogeneity to maintaining glucose homeostasis. Understanding the function and

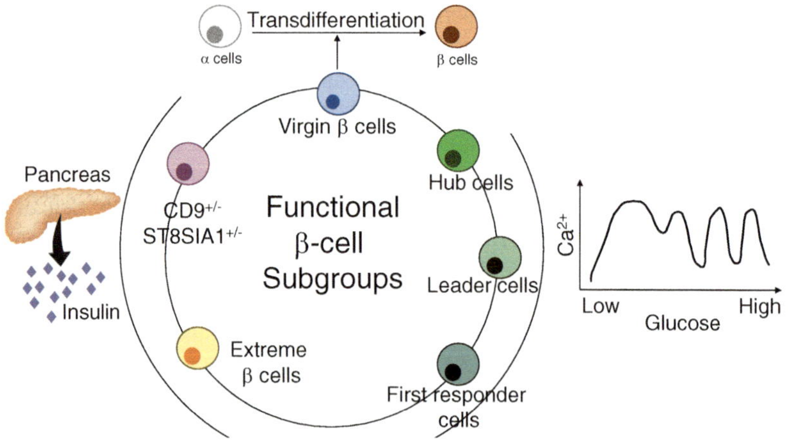

Fig. 3 Functional β-cell subgroups. Virgin β cells are the intermediate stage of transdifferentiation from α to β cells. Hub cells, leader cells, and first-responder cells contribute to $[Ca^{2+}]$ oscillations. Extreme β cells and $CD9^{+/-}ST8SIA1^{+/-}$ cells are involved in basal insulin secretion

cooperation of various β-cell subgroups in regulating glucose levels may help to further the development of innovative diabetes treatments.

10 Virgin β Cells

A population of immature β cells, called virgin β cells, was discovered in a specific environment in the mouse islet periphery. They comprise 1–2% of adult mouse β cells (van der Meulen et al. 2017). Virgin β cells express insulin but do not express key β-cell maturity markers, including UCN3. The absence of these maturity markers and certain functional genes leads to the failure of virgin β cells to participate in glucose-stimulated coordinated pulsatile $[Ca^{2+}]$ oscillations; this subpopulation also exhibits low glucose uptake. Moreover, virgin β cells represent an intermediate stage in the transdifferentiation from α to β cells (van der Meulen et al. 2017). Since β-cell proliferation peaks in the neonatal period, immature virgin β cells could logically serve as a source of β-cell-mass maintenance and be characterized by elevated cell division. However, basal virgin β-cell replication did not differ from the general β-cell population; in addition, their proliferation declines with age. Transition from virgin β-cell to mature β-cell state was rare (Lee et al. 2021). The existence of virgin β cells in humans is still unclear. Therefore, the role of virgin β cells with unique properties in healthy and diabetic humans remains unresolved.

11 Hub Cells

β-cell subpopulations with coordinated pulsatile $[Ca^{2+}]$ oscillations have been identified (Johnston et al. 2016; Westacott et al. 2017; Salem et al. 2019; Kravets et al. 2022). β cells respond to glucose with biphasic insulin secretion: a sharp first-phase peak and a prolonged, lower second-phase peak. According to the limited pool model, the first phase occurs when β cells release insulin granules docked in the membrane in response to increased glucose (Porte Jr. and Pupo 1969; Henquin 1994; Nesher and Cerasi 2002; Rorsman and Renström 2003; Straub and Sharp 2004). The second phase occurs when β cells continue to transport and release newly synthesized insulin secretory granules (Porte Jr. and Pupo 1969; Curry et al. 1968; Komatsu et al. 2001; Aizawa et al. 2002; Porksen 2002; Straub and Sharp 2002). $[Ca^{2+}]$, which drives insulin secretion, fluctuates in response to elevated glucose, and drives insulin release (Wojtusciszyn et al. 2008; Nunemaker et al. 2006). According to a second, conflicting model, termed the limited signaling model, signaling events alternatively potentiate and inhibit insulin release, dependent on the difference between basal and elevated glucose (Caumo and Luzi 2004). That $[Ca^{2+}]$ has a first- and second-wave response to changing glucose levels and controls insulin release supports this second model.

Recently, multiple groups have attempted to define a "pacemaker" cell from whence coordinated signaling originates. One islet functional architecture study found that ~10% of β cells, termed hub cells, display the most physical links to other β cells (Johnston et al. 2016). Hub cells are transcriptionally immature but highly metabolically active. They regulate insulin secretion upon glucose stimuli by setting the pace of electrical dynamics in other β cells. Contact with multiple other β cells allows hub cells to set the pace of electrical dynamics to coordinate [Ca^{2+}] flux throughout the islet. Silencing of hub cells impairs [Ca^{2+}] oscillations across islets, whereas silencing of non-hub cells has no significant effect.

In mice, other β-cell subpopulations that disproportionally control islet electrical dynamics were determined using optogenetics, imaging, and a computational model (Westacott et al. 2017). One of the β-cell subpopulations was metabolically active and governed the electrical response of neighboring β cells. Another population was metabolically less active but propagated and synchronized [Ca^{2+}] oscillations (Westacott et al. 2017).

In zebrafish *Danio rerio*, the [Ca^{2+}] wave originates from temporally defined leader β cells that respond first to glucose stimulation (Salem et al. 2019). Photoablation of leader β cells altered [Ca^{2+}] dynamics in the remaining β cells. Murine islets engrafted into the anterior eye chamber revealed a robust connectivity in a similar β-cell population to that observed in zebrafish. Single-cell RNA sequencing confirmed this β-cell population comprised 9% and 11% of total β cells in zebrafish and mice, respectively (Salem et al. 2019). This β-cell subpopulation identified in zebrafish and mouse islets in vivo resembles the hub cells that were previously defined in vitro (Johnston et al. 2016).

In isolated mouse islets, a β-cell population was identified that responded the earliest to glucose stimulation with elevated [Ca^{2+}] (Kravets et al. 2022). High membrane excitability but depressed electric coupling compared to other β cells characterizes these first-responder cells. Ablation of these first-responder cells decreased [Ca^{2+}] elevation and impaired the first phase of insulin secretion (Kravets et al. 2022). All β-cell subpopulations mentioned in this section control [Ca^{2+}] oscillations to regulate insulin secretion in response to glucose stimulation. However, they differ in metabolic activity, engagement with the first or second wave of [Ca^{2+}] elevation, connectivity, and electrical coupling to neighboring cells. Whether or not the subpopulations addressed above are, in fact, the same cell subgroup or are distinct subpopulations has not been resolved.

12 Extreme β Cells

Single-molecule fluorescence in situ hybridization (smFISH) in intact mouse islets revealed extreme β cells located within the islet core (Farack et al. 2019). Extreme β cells express high levels of ribosomal RNA and mRNA related to insulin secretion, but in the fasting state, they express less mature insulin protein (Farack et al. 2019). Based on these unique properties, extreme β cells are regarded as a β-cell subgroup responsible for basal insulin secretion (Farack et al. 2019). In addition, the proportion of extreme β cells increased in *Lepr*$^{db/db}$ mice, supporting their potential role in basal insulin secretion (Farack et al. 2019; Hudish et al. 2019). Another study applying a modified smFISH to mouse pancreata confirmed the existence of a β-cell population in the islet core with a higher ratio of *NK2 homeobox 2* (*Nkx2.2*) to *Ins2* mRNA (Cui et al. 2018). However, further investigations are needed to identify whether these cells are the same as those other groups classified as extreme β cells.

13 CD9 and ST8SIA1

CD9 is a tetraspanin protein involved in multiple biological processes, including protein trafficking, membrane fusion, and adhesion (Hemler 2005; Andreu and Yanez-Mo 2014). CD9 has been suggested as a negative marker of functional β-like cells derived from human pluripotent stem cells (Li et al. 2020). In contrast, the function of ST8 α-N-acetyl-neuraminide α-2,8-sialyltransferase 1 (ST8SIA1), which catalyzes the transfer of sialic acid from CMP-sialic acid to monosialodihexosylganglioside (GM2), is unknown in β cells (Meter et al. 2022). Human β cells have been classified into four subpopulations based on the differential expression of CD9 and ST8SIA1 (Dorrell et al. 2016). These subpopulations show differential gene expression profiles and GSIS function. In nondiabetic subjects, less functional $CD9^-/ST8SIA1^-$ β cells were the most abundant subpopulation but exhibited the lowest basal insulin secretion and the highest stimulation index (SI). Conversely, $CD9^+/ST8SIA1^+$ β cells were the rarest population but showed the most elevated basal insulin secretion and the lowest SI. In T2D patients, the ratio of $CD9^-/ST8SIA1^-$ β cells substantially decreased while the proportion of $CD9^+/ST8SIA1^+$ β cells significantly increased (Dorrell et al. 2016). Thus, future studies about the kinetics of the four β-cell subtypes isolated by the cell surface markers CD9 and ST8SIA1 may provide insight into T2D progression and their cooperation in insulin secretion.

14 Other Islet Cell Plasticity

Islet cells with transitional states are understudied in the literature because they are rare, so they are easily missed in bulk studies. Furthermore, they can resemble doublets in gene expressions, so they are generally removed in single-cell RNA-sequencing analysis. Nevertheless, recent findings have reported transdifferentiation from exocrine and endocrine cells to β cells. A larger number of double-positive cells, such as cytokeratin $(CK)19^+/INS^+$, GCG^+/INS^+, and somatostatin $(SST)^+/INS^+$, were detected in insulin-resistant, nondiabetic subjects. These double-positive cells were scattered throughout the pancreas rather than localized to large islets (Yoneda et al. 2013; Mezza et al. 2014). In mice, β cells co-expressing multiple islet hormone genes were identified by nested RT-PCR in adults, embryos, and neonates (Katsuta et al. 2010). As mentioned in the dedifferentiation/transdifferentiation section above, some of the dedifferentiated progenitor-like β cells converted to α- or δ-like cells in T2D (Cinti et al. 2016). In this section, we focus on the plasticity of non-β pancreatic cells and their promising contribution to β-cell regeneration.

15 Endocrine Cells

Severe β-cell loss can trigger transdifferentiation from α and δ cells to insulin-producing bi-hormonal cells (Zhao et al. 2021). In addition, α cells can convert to β cells from puberty through old age after a nearly complete β-cell ablation in mice. However, this conversion was not detected before puberty. In contrast, transdifferentiation from δ cells to β cells was observed in juvenile mice after β-cell ablation. These studies imply endocrine cell plasticity after β-cell loss in an age-dependent manner (Chera et al. 2014).

α cells reciprocally control blood glucose with β cells by secreting glucagon to increase blood glucose levels by inducing glycogenolysis. α cells are developmentally closer to β cells than exocrine cells, and overlapping transcription profiles between α and β cells support this relationship (Baeyens et al. 2018; Wang et al. 2016a; Segerstolpe et al. 2016; Lawlor et al. 2017). In diabetes, α cells are more resistant to metabolic stress than β cells and proliferate more frequently than β cells (Wang et al. 2016b; Marroqui et al. 2015; Lam et al. 2018). Lineage-tracing studies have revealed that β cells regenerated from α cells after complete β-cell ablation (Thorel et al. 2010). These results raise the potential role of α-cell plasticity in β-cell mass homeostasis after β-cell loss.

Remarkably, several pieces of evidence support the epigenomic plasticity of α cells to β cells. Chromatin immunoprecipitation sequencing (ChIP-seq) of both active H3K4me3 and repressive H3K27me3 histone marks on purified human α and β cells identified a bivalent chromatin state in α cells. The bivalent marks in α cells resemble those in embryonic stem cells, implying a multipotent epigenomic state of α cells (Bramswig et al. 2013). Assay for transposase-accessible chromatin with sequencing (ATAC-seq) confirmed this finding and showed that α cells exhibit open chromatin feature in β-cell selective markers (Ackermann et al. 2016). These discoveries imply that α cells are epigenetically poised to convert to β cells. Supporting this hypothesis, the inhibition of histone methyltransferase reduces the levels of repressive histone marks and induces the co-expression of glucagon and insulin in human and mouse islet cells (Bramswig et al. 2013). In mice, inhibition of DNA methyltransferase 1 (DNMT1) and aristaless-related homeobox (ARX) induces transdifferentiation from α cells to β cells (Chakravarthy et al. 2017). Single-cell RNA sequencing revealed little transcriptional difference between converted β cells and bona fide β cells (Chakravarthy et al. 2017). T1D donor pancreata also contained α cells expressing insulin and low levels of ARX and DNMT1 (Chakravarthy et al. 2017). Further, several groups reported the upregulation of disallowed markers in α cells in T1D pathology (Brissova et al. 2018; Wang et al. 2019). These findings suggest unexplored human α- and β-cell plasticity in T1D. Several other models of α- to β-cell conversion have been reported in animal models, both through genetic and small-molecule intervention (Ye et al. 2015; Ben-Othman et al. 2017). For example, overexpression of ectopic paired box 4 (PAX4) converted α cells to β cells (Collombat et al. 2009; Al-Hasani et al. 2013).

Recently, a rare NEUROG3$^+$ ε-cell population was reported in the postnatal human pancreas, and the transcriptional similarities between postnatal NEUROG3$^+$ cells and NEUROG3$^+$ cells in the fetal pancreas were confirmed (Yong et al. 2021). Since NEUROG3 is a well-known master regulator of pancreatic endocrine differentiation, this result suggests the possibility of NEUROG3$^+$ ε cells converting to other endocrine cell types (Gradwohl et al. 2000; D'Amour et al. 2006; Pagliuca et al. 2014). This result is supported by studies showing that ε cells gave rise to α, PP, and rare β cells in mice (Arnes et al. 2012; Yu et al. 2021). Knowledge gained about non-β endocrine cell plasticity will aid in developing β-cell regeneration therapies to cure diabetes.

16 Exocrine Cells

Acinar and ductal cells are pancreatic exocrine cells that produce and transport digestive enzymes through the pancreas to the duodenum. Several single-cell RNA-sequencing studies and a single-nucleus RNA-sequencing approach on human islets discovered acinar cell heterogeneity (Muraro et al. 2016; Segerstolpe et al. 2016; Wollny et al. 2016; Tosti et al. 2021). Segerstolpe et al. split acinar cells into two clusters and found one group contained more proliferative acinar cells (Segerstolpe et al. 2016). Muraro et al. identified four clusters of acinar cells, including a *regenerating family member 3 α* (*REG3A*)$^+$/*serine protease 1* (*PRSS1*)^{low} acinar cell cluster. Immunostaining further confirmed the presence of many *REG3A+*/*PRSS1*^{low} acinar cells in the human pancreas (Muraro et al. 2016). Wollny et al. suggested pancreatic injury as a potential trigger for progenitor-like acinar cells to differentiate into β cells. While most acinar cells showed limited proliferation, the progenitor-like acinar cell population had a long-term proliferative capacity and expressed progenitor markers, including stathmin 1 (STMN1). Remarkably, proliferation and expression of STMN1 were transiently upregulated in acinar cells upon cerulein-induced injury leading to acute pancreatitis (Wollny et al. 2016). Tosti et al. uncovered three acinar cell clusters. The first cluster was characterized by higher expression of *REG1B*, *REG3A*, and *REG3G*, similar to previous reports (Muraro et al. 2016; Tosti et al. 2021). The remaining two acinar cell clusters featured two distinct digestive-enzyme-related gene sets (Tosti et al. 2021). However, any potential contribution of these acinar cells to β-cell regeneration needs to be further studied.

On the other hand, reprogramming of acinar cells in different settings has demonstrated the plasticity of acinar cells (Baeyens et al. 2005; Minami et al. 2005). In mice, the conversion from acinar to β cells was induced by combinatorial expression of NEUROG3, PDX1, and MAFA in vivo. The converted β cells expressed typical β cell markers and regulated glucose homeostasis in diabetic mice, but the β cells did not aggregate into islets (Zhou et al. 2008). Additionally, transdifferentiation from human acinar cells to β cells via viral expression of mitogen-activated protein kinase (MAPK) and signal transducer and activator of transcription 3 (STAT3) was shown in a mouse xenotransplantation model (Lemper et al. 2015). In adult mice under pancreatic duct ligation, pancreas transcription factor 1 subunit α $(PTF1a)^+$ acinar cells transdifferentiate to ductal cells and eventually to β cells (Pan et al. 2013).

Duct-cell heterogeneity has also been investigated by single-cell RNA sequencing and single-nucleus RNA sequencing on human islets (Baron et al. 2016; Tosti et al. 2021). Both groups found two distinct ductal cell populations. Baron et al. demonstrated that the first cluster highly expressed *trefoil factor 1 (TFF1)*, *TFF2*, and *mucin 1 (MUC1)*, suggesting this population secretes mucus to protect duct cells from digestive enzymes. The other cluster highly expressed cystic fibrosis transmembrane conductance regulator (CFTR), a gene associated with the secretory function of ductal cells. Interestingly, two clusters were also spatially separated by immunostaining of pancreatic tissue (Baron et al. 2016). The two distinct ductal populations discovered by Tosti et al. were very similar to those found by Baron et al. The first cluster was characterized by higher expression of *TFF1*, *TFF2*, *TFF3*, and *MUC5B*. The other cluster featured higher expression of *CFTR*, *secretin receptor* (*SCTR*), and *solute carrier family 4 member 4* (*SLC4A4*), as revealed in the study above (Baron et al. 2016; Tosti et al. 2021). However, the role of these heterogeneous clusters has not yet been elucidated.

Ductal cell conversion to β cells has been studied extensively but remains controversial. In neonatal rat pancreas, new β cells found in the islet periphery were derived from cells expressing the ductal cell markers CK17 and CK19 (Bouwens et al. 1994). However, a contradictory study argued that duct cells give rise to β cells only during embryogenesis (Solar et al. 2009). Some studies have suggested that ductal cells also converted to β cells during pregnancy or after injury. In pregnant women, an increase in scattered β cells and insulin$^+$ ductal cells was observed (Butler et al. 2010). Ductal cells expressing GLUT2 and β cells derived from NEUROG3$^+$ ductal cells were observed upon injury (Xu et al. 2008; Wang et al. 1995). However, a contrasting study reported that mouse ductal cells expressing SOX9, a developmental marker, did not give rise to β cells upon injury (Kopp et al. 2011). Therefore, further studies are required to confirm whether ductal cells can contribute to β-cell expansion upon injury and during pregnancy in humans.

Like acinar cells, in mice, ductal cells can be converted into β cells by inducing *Neurog3*, *Pdx1*, and *MafA* expression (Al-Hasani et al. 2013; Yamada et al. 2015). In addition, ductal cells from digested human pancreatic tissue cultured in conditioned media expanded and gave rise to human islets, including insulin-producing β cells (Bonner-Weir et al. 2000). This experiment demonstrates the remarkable plasticity of duct cells. Thus, given exocrine cells compose most of the pancreatic mass, acinar and ductal cells are appealing sources to regenerate β cells. However, further studies are needed to develop effective strategies to reprogram human ductal cells in vitro to regenerate massive amounts of β cells for curing diabetes.

17 Conclusion

Advanced technologies have furthered knowledge regarding β-cell heterogeneity and plasticity. β-cell stress, such as aging and ER stress, leads to β-cell heterogeneity and enforces the plasticity of β cells for adaptation to environmental changes. To maintain glucose and β-cell homeostasis, β cells manifest flexibility in proliferation, dedifferentiation or transdifferentiation, maturation, and insulin secretion. Functional β-cell subgroups exist that differ in maturity, electrical activity, and insulin secretion. Furthermore, non-β pancreatic cells may convert to β cells to maintain β-cell homeostasis.

Various β-cell populations exist simultaneously, but the characteristics of individual β cells can gradually change, depending on time and environment. Although newly developed computational algorithms can help infer cellular transitional processes from single-cell RNA-sequencing data (Trapnell et al. 2014; La Manno et al. 2018), these approaches lack the temporal resolution required to understand population dynamics. They also rely on several assumptions, such as a smooth and continuous transition between different cellular states, which may not be a trait of every biological system. Therefore, to apply this knowledge to develop effective diabetes therapies, tools to elucidate the dynamics of human β-cell subpopulations and measure their molecular and functional differences during diabetes progression need to be developed. Overall, further studies on the dynamics of human β-cell subpopulations and their molecular and functional differences during diabetes progression will aid in understanding β-cell biology and developing novel diabetes treatments.

Acknowledgments This work was supported by YJW's grants from the JDRF (1-INO-2022-1129-A-N) and the Helmsley Charitable Trust George Eisenbarth nPOD Award for Team Science (2018PG-T1D060). We thank the authors of previous β-cell heterogeneity and plasticity studies.

Compliance with Ethical Standards Hyo Jeong Yong declares no conflicts of interest. Yue J. Wang declares no conflicts of interest. This chapter does not contain any studies with any human participants or animals performed by any of the authors.

References

Abdul-Rasoul M, Habib H, Al-Khouly M (2006) 'The honeymoon phase' in children with type 1 diabetes mellitus: frequency, duration, and influential factors. Pediatr Diabetes 7(2):101–107

Ackermann Misfeldt A, Costa RH, Gannon M (2008) Beta-cell proliferation, but not neogenesis, following 60% partial pancreatectomy is impaired in the absence of FoxM1. Diabetes 57(11):3069–3077

Ackermann AM, Wang Z, Schug J, Naji A, Kaestner KH (2016) Integration of ATAC-seq and RNA-seq identifies human alpha cell and beta cell signature genes. Mol Metab 5(3):233–244
Afelik S, Rovira M (2017) Pancreatic beta-cell regeneration: facultative or dedicated progenitors? Mol Cell Endocrinol 445:85–94
Aguayo-Mazzucato C, van Haaren M, Mruk M, Lee TB Jr, Crawford C, Hollister-Lock J, Sullivan BA, Johnson JW, Ebrahimi A, Dreyfuss JM, Van Deursen J, Weir GC, Bonner-Weir S (2017) Beta cell aging markers have heterogeneous distribution and are induced by insulin resistance. Cell Metab 25(4):898–910 e895
Aizawa T, Sato Y, Komatsu M (2002) Importance of nonionic signals for glucose-induced biphasic insulin secretion. Diabetes 51(Suppl 1):S96–S98
Alarcon C, Boland BB, Uchizono Y, Moore PC, Peterson B, Rajan S, Rhodes OS, Noske AB, Haataja L, Arvan P, Marsh BJ, Austin J, Rhodes CJ (2016) Pancreatic beta-cell adaptive plasticity in obesity increases insulin production but adversely affects secretory function. Diabetes 65(2):438–450
Al-Hasani K, Pfeifer A, Courtney M, Ben-Othman N, Gjernes E, Vieira A, Druelle N, Avolio F, Ravassard P, Leuckx G, Lacas-Gervais S, Ambrosetti D, Benizri E, Hecksher-Sorensen J, Gounon P, Ferrer J, Gradwohl G, Heimberg H, Mansouri A, Collombat P (2013) Adult duct-lining cells can reprogram into beta-like cells able to counter repeated cycles of toxin-induced diabetes. Dev Cell 26(1):86–100
Anderson JW, Kendall CWC, Jenkins DJA (2003) Importance of weight management in type 2 diabetes: review with meta-analysis of clinical studies. J Am Coll Nutr 22(5):331–339
Andreu Z, Yanez-Mo M (2014) Tetraspanins in extracellular vesicle formation and function. Front Immunol 5:442
Arnes L, Hill JT, Gross S, Magnuson MA, Sussel L (2012) Ghrelin expression in the mouse pancreas defines a unique multipotent progenitor population. PLoS One 7(12):e52026
Arrojo EDR (2021) Probing beta-cell biology in space and time. Diabetes 70(10):2163–2173
Arrojo EDR, Lev-Ram V, Tyagi S, Ramachandra R, Deerinck T, Bushong E, Phan S, Orphan V, Lechene C, Ellisman MH, Hetzer MW (2019) Age mosaicism across multiple scales in adult tissues. Cell Metab 30(2):343–351 e343
Avrahami D, Wang YJ, Schug J, Feleke E, Gao L, Liu C, Consortium H, Naji A, Glaser B, Kaestner KH (2020) Single-cell transcriptomics of human islet ontogeny defines the molecular basis of β-cell dedifferentiation in T2D. Mol Metab 42:101057
Back SH, Scheuner D, Han J, Song B, Ribick M, Wang J, Gildersleeve RD, Pennathur S, Kaufman RJ (2009) Translation attenuation through eIF2α phosphorylation prevents oxidative stress and maintains the differentiated state in β cells. Cell Metab 10(1):13–26
Baeyens L, De Breuck S, Lardon J, Mfopou JK, Rooman I, Bouwens L (2005) In vitro generation of insulin-producing beta cells from adult exocrine pancreatic cells. Diabetologia 48(1):49–57
Baeyens L, Lemper M, Staels W, De Groef S, De Leu N, Heremans Y, German MS, Heimberg H (2018) (Re)generating human beta cells: status, pitfalls, and perspectives. Physiol Rev 98(3):1143–1167
Baron M, Veres A, Wolock SL, Faust AL, Gaujoux R, Vetere A, Ryu JH, Wagner BK, Shen-Orr SS, Klein AM, Melton DA, Yanai I (2016) A single-cell transcriptomic map of the human and mouse pancreas reveals inter- and intra-cell population structure. Cell Syst 3(4):346–360 e344
Bengtsson M, Stahlberg A, Rorsman P, Kubista M (2005) Gene expression profiling in single cells from the pancreatic islets of Langerhans reveals lognormal distribution of mRNA levels. Genome Res 15(10):1388–1392
Ben-Othman N, Vieira A, Courtney M, Record F, Gjernes E, Avolio F, Hadzic B, Druelle N, Napolitano T, Navarro-Sanz S, Silvano S, Al-Hasani K, Pfeifer A, Lacas-Gervais S, Leuckx G, Marroqui L, Thevenet J, Madsen OD, Eizirik DL, Heimberg H, Kerr-Conte J, Pattou F, Mansouri A, Collombat P (2017) Long-term GABA administration induces alpha cell-mediated beta-like cell neogenesis. Cell 168(1–2):73–85 e11
Bernales S, Papa FR, Walter P (2006) Intracellular signaling by the unfolded protein response. Annu Rev Cell Dev Biol 22:487–508

Block MB, Rosenfield RL, Mako ME, Steiner DF, Rubenstein AH (1973) Sequential changes in beta-cell function in insulin-treated diabetic patients assessed by C-peptide immunoreactivity. N Engl J Med 288(22):1144–1148

Bonner-Weir S, Taneja M, Weir GC, Tatarkiewicz K, Song KH, Sharma A, O'Neil JJ (2000) In vitro cultivation of human islets from expanded ductal tissue. Proc Natl Acad Sci USA 97(14):7999–8004

Bouwens L, Wang RN, De Blay E, Pipeleers DG, Kloppel G (1994) Cytokeratins as markers of ductal cell differentiation and islet neogenesis in the neonatal rat pancreas. Diabetes 43(11):1279–1283

Bramswig NC, Everett LJ, Schug J, Dorrell C, Liu C, Luo Y, Streeter PR, Naji A, Grompe M, Kaestner KH (2013) Epigenomic plasticity enables human pancreatic alpha to beta cell reprogramming. J Clin Invest 123(3):1275–1284

Brandhorst H, Brandhorst D, Hering BJ, Federlin K, Bretzel RG (1995) Body mass index of pancreatic donors: a decisive factor for human islet isolation. Exp Clin Endocrinol Diabetes 103(Suppl 2):23–26

Brereton MF, Iberl M, Shimomura K, Zhang Q, Adriaenssens AE, Proks P, Spiliotis II, Dace W, Mattis KK, Ramracheya R, Gribble FM, Reimann F, Clark A, Rorsman P, Ashcroft FM (2014) Reversible changes in pancreatic islet structure and function produced by elevated blood glucose. Nat Commun 5(1):4639

Brissova M, Haliyur R, Saunders D, Shrestha S, Dai C, Blodgett DM, Bottino R, Campbell-Thompson M, Aramandla R, Poffenberger G, Lindner J, Pan FC, von Herrath MG, Greiner DL, Shultz LD, Sanyoura M, Philipson LH, Atkinson M, Harlan DM, Levy SE, Prasad N, Stein R, Powers AC (2018) α cell function and gene expression are compromised in type 1 diabetes. Cell Rep 22(10):2667–2676

Brunzell JD, Robertson RP, Lerner RL, Hazzard WR, Ensinck JW, Bierman EL, Porte D Jr (1976) Relationships between fasting plasma glucose levels and insulin secretion during intravenous glucose tolerance tests. J Clin Endocrinol Metab 42(2):222–229

Brush JM (1944) Initial stabilization of the diabetic child. Am J Dis Child 67(6):429–444

Butler AE, Janson J, Bonner-Weir S, Ritzel R, Rizza RA, Butler PC (2003) Beta-cell deficit and increased beta-cell apoptosis in humans with type 2 diabetes. Diabetes 52(1):102–110

Butler AE, Galasso R, Meier JJ, Basu R, Rizza RA, Butler PC (2007) Modestly increased beta cell apoptosis but no increased beta cell replication in recent-onset type 1 diabetic patients who died of diabetic ketoacidosis. Diabetologia 50(11):2323–2331

Butler AE, Cao-Minh L, Galasso R, Rizza RA, Corradin A, Cobelli C, Butler PC (2010) Adaptive changes in pancreatic beta cell fractional area and beta cell turnover in human pregnancy. Diabetologia 53(10):2167–2176

Camastra S, Manco M, Mari A, Baldi S, Gastaldelli A, Greco AV, Mingrone G, Ferrannini E (2005) Beta-cell function in morbidly obese subjects during free living: long-term effects of weight loss. Diabetes 54(8):2382–2389

Campbell-Thompson M, Fu A, Kaddis JS, Wasserfall C, Schatz DA, Pugliese A, Atkinson MA (2016) Insulitis and beta-cell mass in the natural history of type 1 diabetes. Diabetes 65(3):719–731

Camunas-Soler J, Dai XQ, Hang Y, Bautista A, Lyon J, Suzuki K, Kim SK, Quake SR, MacDonald PE (2020) Patch-seq links single-cell transcriptomes to human islet dysfunction in diabetes. Cell Metab 31(5):1017–1031 e1014

Cano DA, Rulifson IC, Heiser PW, Swigart LB, Pelengaris S, German M, Evan GI, Bluestone JA, Hebrok M (2008) Regulated beta-cell regeneration in the adult mouse pancreas. Diabetes 57(4):958–966

Casimiro I, Stull ND, Tersey SA, Mirmira RG (2021) Phenotypic sexual dimorphism in response to dietary fat manipulation in C57BL/6J mice. J Diabetes Complicat 35(2):107795

Caumo A, Luzi L (2004) First-phase insulin secretion: does it exist in real life? Considerations on shape and function. Am J Physiol Endocrinol Metab 287(3):E371–E385

Chakravarthy H, Gu X, Enge M, Dai X, Wang Y, Damond N, Downie C, Liu K, Wang J, Xing Y, Chera S, Thorel F, Quake S, Oberholzer J, MacDonald PE, Herrera PL, Kim SK (2017)

Converting adult pancreatic islet alpha cells into beta cells by targeting both Dnmt1 and Arx. Cell Metab 25(3):622–634
Chang AM, Halter JB (2003) Aging and insulin secretion. Am J Physiol Endocrinol Metab 284(1):E7–E12
Chen M, Bergman RN, Pacini G, Porte D Jr (1985) Pathogenesis of age-related glucose intolerance in man: insulin resistance and decreased beta-cell function. J Clin Endocrinol Metab 60(1):13–20
Chen C, Cohrs CM, Stertmann J, Bozsak R, Speier S (2017) Human beta cell mass and function in diabetes: recent advances in knowledge and technologies to understand disease pathogenesis. Mol Metab 6(9):943–957
Chen C-W, Guan B-J, Alzahrani MR, Gao Z, Gao L, Bracey S, Haataja L, Schaffer AE, Lee H, Laframboise T, Bederman I, Arvan P, Mathews CE, Gerling IC, Kaestner KH, Tirosh B, Engin F, Hatzoglou M (2022) Adaptation to chronic ER stress enforces pancreatic β-cell plasticity. Nat Commun 13(1):4621
Chera S, Baronnier D, Ghila L, Cigliola V, Jensen JN, Gu G, Furuyama K, Thorel F, Gribble FM, Reimann F, Herrera PL (2014) Diabetes recovery by age-dependent conversion of pancreatic δ-cells into insulin producers. Nature 514(7523):503–507
Chmelova H, Cohrs CM, Chouinard JA, Petzold C, Kuhn M, Chen C, Roeder I, Kretschmer K, Speier S (2015) Distinct roles of beta-cell mass and function during type 1 diabetes onset and remission. Diabetes 64(6):2148–2160
Cho YS, Chen C-H, Hu C, Long J, RTH O, Sim X, Takeuchi F, Wu Y, Go MJ, Yamauchi T, Chang Y-C, Kwak SH, RCW M, Yamamoto K, Adair LS, Aung T, Cai Q, Chang L-C, Chen Y-T, Gao Y, Hu FB, Kim H-L, Kim S, Kim YJ, Lee JJ-M, Lee NR, Li Y, Liu JJ, Lu W, Nakamura J, Nakashima E, Ng DP-K, Tay WT, Tsai F-J, Wong TY, Yokota M, Zheng W, Zhang R, Wang C, So WY, Ohnaka K, Ikegami H, Hara K, Cho YM, Cho NH, Chang T-J, Bao Y, Hedman ÅK, Morris AP, MI MC, Consortium D, Mu TC, Takayanagi R, Park KS, Jia W, Chuang L-M, JCN C, Maeda S, Kadowaki T, Lee J-Y, Wu J-Y, Teo YY, Tai ES, Shu XO, Mohlke KL, Kato N, Han B-G, Seielstad M (2011) Meta-analysis of genome-wide association studies identifies eight new loci for type 2 diabetes in east Asians. Nat Genet 44(1):67–72
Chu WS, Das SK, Wang H, Chan JC, Deloukas P, Froguel P, Baier LJ, Jia W, McCarthy MI, Ng MCY, Damcott C, Shuldiner AR, Zeggini E, Elbein SC (2007) Activating transcription factor 6 (ATF6) sequence polymorphisms in type 2 diabetes and pre-diabetic traits. Diabetes 56(3):856–862
Cinti F, Bouchi R, Kim-Muller JY, Ohmura Y, Sandoval PR, Masini M, Marselli L, Suleiman M, Ratner LE, Marchetti P, Accili D (2016) Evidence of beta-cell dedifferentiation in human type 2 diabetes. J Clin Endocrinol Metab 101(3):1044–1054
Clark A, Wells CA, Buley ID, Cruickshank JK, Vanhegan RI, Matthews DR, Cooper GJ, Holman RR, Turner RC (1988) Islet amyloid, increased A-cells, reduced B-cells and exocrine fibrosis: quantitative changes in the pancreas in type 2 diabetes. Diabetes Res 9(4):151–159
Cnop M, Hughes SJ, Igoillo-Esteve M, Hoppa MB, Sayyed F, van de Laar L, Gunter JH, de Koning EJP, Walls GV, Gray DWG, Johnson PRV, Hansen BC, Morris JF, Pipeleers-Marichal M, Cnop I, Clark A (2010) The long lifespan and low turnover of human islet beta cells estimated by mathematical modelling of lipofuscin accumulation. Diabetologia 53(2):321–330
Collombat P, Xu X, Ravassard P, Sosa-Pineda B, Dussaud S, Billestrup N, Madsen OD, Serup P, Heimberg H, Mansouri A (2009) The ectopic expression of Pax4 in the mouse pancreas converts progenitor cells into alpha and subsequently beta cells. Cell 138(3):449–462
Criscimanna A, Speicher JA, Houshmand G, Shiota C, Prasadan K, Ji B, Logsdon CD, Gittes GK, Esni F (2011) Duct cells contribute to regeneration of endocrine and acinar cells following pancreatic damage in adult mice. Gastroenterology 141(4):1451–1462. 1462.e1451-1456
Cui Y, Hu D, Markillie LM, Chrisler WB, Gaffrey MJ, Ansong C, Sussel L, Orr G (2018) Fluctuation localization imaging-based fluorescence in situ hybridization (fliFISH) for accurate detection and counting of RNA copies in single cells. Nucleic Acids Res 46(2):e7
Curry DL, Bennett LL, Grodsky GM (1968) Dynamics of insulin secretion by the perfused rat pancreas. Endocrinology 83(3):572–584

D'Amour KA, Bang AG, Eliazer S, Kelly OG, Agulnick AD, Smart NG, Moorman MA, Kroon E, Carpenter MK, Baetge EE (2006) Production of pancreatic hormone–expressing endocrine cells from human embryonic stem cells. Nat Biotechnol 24(11):1392–1401

De Tata V (2014) Age-related impairment of pancreatic beta-cell function: pathophysiological and cellular mechanisms. Front Endocrinol 5:138

Deng S, Vatamaniuk M, Huang X, Doliba N, Lian MM, Frank A, Velidedeoglu E, Desai NM, Koeberlein B, Wolf B, Barker CF, Naji A, Matschinsky FM, Markmann JF (2004) Structural and functional abnormalities in the islets isolated from type 2 diabetic subjects. Diabetes 53(3):624–632

Dor Y, Brown J, Martinez OI, Melton DA (2004) Adult pancreatic beta-cells are formed by self-duplication rather than stem-cell differentiation. Nature 429(6987):41–46

Dorrell C, Schug J, Canaday PS, Russ HA, Tarlow BD, Grompe MT, Horton T, Hebrok M, Streeter PR, Kaestner KH, Grompe M (2016) Human islets contain four distinct subtypes of beta cells. Nat Commun 7:11756

Enge M, Arda HE, Mignardi M, Beausang J, Bottino R, Kim SK, Quake SR (2017) Single-cell analysis of human pancreas reveals transcriptional signatures of aging and somatic mutation patterns. Cell 171(2):321–330 e314

Engin F, Nguyen T, Yermalovich A, Hotamisligil GS (2014) Aberrant islet unfolded protein response in type 2 diabetes. Sci Rep 4:4054

Ernst S, Demirci C, Valle S, Velazquez-Garcia S, Garcia-Ocana A (2011) Mechanisms in the adaptation of maternal beta-cells during pregnancy. Diabetes Manag 1(2):239–248

Farack L, Golan M, Egozi A, Dezorella N, Bahar Halpern K, Ben-Moshe S, Garzilli I, Toth B, Roitman L, Krizhanovsky V, Itzkovitz S (2019) Transcriptional heterogeneity of beta cells in the intact pancreas. Dev Cell 48(1):115–125 e114

Ferrannini E, Natali A, Bell P, Cavallo-Perin P, Lalic N, Mingrone G (1997) Insulin resistance and hypersecretion in obesity. European Group for the Study of Insulin Resistance (EGIR). J Clin Invest 100(5):1166–1173

Ferrannini E, Gastaldelli A, Miyazaki Y, Matsuda M, Mari A, DeFronzo RA (2005) Beta-cell function in subjects spanning the range from normal glucose tolerance to overt diabetes: a new analysis. J Clin Endocrinol Metab 90(1):493–500

Fonolleda M, Murillo M, Vázquez F, Bel J, Vives-Pi M (2017) Remission phase in paediatric type 1 diabetes: new understanding and emerging biomarkers. Horm Res Paediatr 88(5):307–315

Freemark M, Avril I, Fleenor D, Driscoll P, Petro A, Opara E, Kendall W, Oden J, Bridges S, Binart N, Breant B, Kelly PA (2002) Targeted deletion of the PRL receptor: effects on islet development, insulin production, and glucose tolerance. Endocrinology 143(4):1378–1385

Gastaldelli A, Ferrannini E, Miyazaki Y, Matsuda M, DeFronzo RA, San Antonio metabolism S (2004) Beta-cell dysfunction and glucose intolerance: results from the San Antonio metabolism (SAM) study. Diabetologia 47(1):31–39

Gepts W (1965) Pathologic anatomy of the pancreas in juvenile diabetes mellitus. Diabetes 14(10):619–633

Gepts W, De Mey J (1978) Islet cell survival determined by morphology. An immunocytochemical study of the islets of Langerhans in juvenile diabetes mellitus. Diabetes 27(Suppl 1):251–261

Gianani R, Putnam A, Still T, Yu L, Miao D, Gill RG, Beilke J, Supon P, Valentine A, Iveson A, Dunn S, Eisenbarth GS, Hutton J, Gottlieb P, Wiseman A (2006) Initial results of screening of nondiabetic organ donors for expression of islet autoantibodies. J Clin Endocrinol Metab 91(5):1855–1861

Giordano E, Bosco D, Cirulli V, Meda P (1991) Repeated glucose stimulation reveals distinct and lasting secretion patterns of individual rat pancreatic B cells. J Clin Invest 87(6):2178–2185

Golson ML, Misfeldt AA, Kopsombut UG, Petersen CP, Gannon M (2010) High fat diet regulation of beta-cell proliferation and beta-cell mass. Open Endocrinol J 4

Gradwohl G, Dierich A, LeMeur M, Guillemot F (2000) Neurogenin3 is required for the development of the four endocrine cell lineages of the pancreas. Proc Natl Acad Sci USA 97(4):1607–1611

Green IC, El Seifi S, Perrin D, Howell SL (1981) Cell replication in the islets of langerhans of adult rats: effects of pregnancy, ovariectomy and treatment with steroid hormones. J Endocrinol https://doi.org/10.1677/joe.0.0880219
Guo S, Dai C, Guo M, Taylor B, Harmon JS, Sander M, Robertson RP, Powers AC, Stein R (2013) Inactivation of specific β cell transcription factors in type 2 diabetes. J Clin Invest 123(8):3305–3316
Gutiérrez GD, Bender AS, Cirulli V, Mastracci TL, Kelly SM, Tsirigos A, Kaestner KH, Sussel L (2017) Pancreatic β cell identity requires continual repression of non–β cell programs. J Clin Invest 127(1):244–259
Hanley SC, Austin E, Assouline-Thomas B, Kapeluto J, Blaichman J, Moosavi M, Petropavlovskaia M, Rosenberg L (2010) {beta}-cell mass dynamics and islet cell plasticity in human type 2 diabetes. Endocrinology 151(4):1462–1472
Heimberg H, De Vos A, Vandercammen A, Van Schaftingen E, Pipeleers D, Schuit F (1993) Heterogeneity in glucose sensitivity among pancreatic beta-cells is correlated to differences in glucose phosphorylation rather than glucose transport. EMBO J 12(7):2873–2879
Hellman B, Gylfe E, Grapengiesser E, Lund PE, Berts A (1992) Cytoplasmic Ca2+ oscillations in pancreatic beta-cells. Biochim Biophys Acta 1113(3–4):295–305
Hemler ME (2005) Tetraspanin functions and associated microdomains. Nat Rev Mol Cell Biol 6(10):801–811
Henquin JC (1994) Cell biology of insulin secretion. Joslin's diabetes mellitus, Boston
Herold KC, Usmani-Brown S, Ghazi T, Lebastchi J, Beam CA, Bellin MD, Ledizet M, Sosenko JM, Krischer JP, Palmer JP (2015) Beta cell death and dysfunction during type 1 diabetes development in at-risk individuals. J Clin Invest 125(3):1163–1173
Hetz C (2012) The unfolded protein response: controlling cell fate decisions under ER stress and beyond. Nat Rev Mol Cell Biol 13(2):89–102
Holman RR (1998) Assessing the potential for alpha-glucosidase inhibitors in prediabetic states. Diabetes Res Clin Pract 40(Suppl):S21–S25
Huang C, Snider F, Cross JC (2009) Prolactin receptor is required for normal glucose homeostasis and modulation of beta-cell mass during pregnancy. Endocrinology 150(4):1618–1626
Hudish LI, Lorberbaum DS, Sussel L (2019) FISHing for β cells. Dev Cell 48(1):7–8
Hull RL, Kodama K, Utzschneider KM, Carr DB, Prigeon RL, Kahn SE (2005) Dietary-fat-induced obesity in mice results in beta cell hyperplasia but not increased insulin release: evidence for specificity of impaired beta cell adaptation. Diabetologia 48(7):1350–1358
Hwang J, Qi L (2018) Quality control in the endoplasmic reticulum: crosstalk between ERAD and UPR pathways. Trends Biochem Sci 43(8):593–605
Ihm SH, Matsumoto I, Sawada T, Nakano M, Zhang HJ, Ansite JD, Sutherland DE, Hering BJ (2006) Effect of donor age on function of isolated human islets. Diabetes 55(5):1361–1368
Ihm SH, Moon HJ, Kang JG, Park CY, Oh KW, Jeong IK, Oh YS, Park SW (2007) Effect of aging on insulin secretory function and expression of beta cell function-related genes of islets. Diabetes Res Clin Pract 77(Suppl 1):S150–S154
In't Veld P, Lievens D, De Grijse J, Ling Z, Van der Auwera B, Pipeleers-Marichal M, Gorus F, Pipeleers D (2007) Screening for insulitis in adult autoantibody-positive organ donors. Diabetes 56(9):2400–2404
Inaishi J, Saisho Y, Sato S, Kou K, Murakami R, Watanabe Y, Kitago M, Kitagawa Y, Yamada T, Itoh H (2016) Effects of obesity and diabetes on alpha- and beta-cell mass in surgically resected human pancreas. J Clin Endocrinol Metab 101(7):2874–2882
Jackson RL, Boyd JD, Smith TE (1940) Stabilization of the diabetic child. Am J Dis Child 59(2):332–341
Jensen CC, Cnop M, Hull RL, Fujimoto WY, Kahn SE, American Diabetes Association GSG (2002) Beta-cell function is a major contributor to oral glucose tolerance in high-risk relatives of four ethnic groups in the U.S. Diabetes 51(7):2170–2178
Johnston NR, Mitchell RK, Haythorne E, Pessoa MP, Semplici F, Ferrer J, Piemonti L, Marchetti P, Bugliani M, Bosco D, Berishvili E, Duncanson P, Watkinson M, Broichhagen J, Trauner D,

Rutter GA, Hodson DJ (2016) Beta cell hubs dictate pancreatic islet responses to glucose. Cell Metab 24(3):389–401

Junker K, Egeberg J, Kromann H, Nerup J (1977) An autopsy study of the islets of Langerhans in acute-onset juvenile diabetes mellitus. Acta Pathol Microbiol Scand A 85(5):699–706

Kaas A, Andersen ML, Fredheim S, Hougaard P, Buschard K, Petersen JS, de Beaufort C, Robertson KJ, Hansen L, Mortensen HB, Nielsen LB (2012) Hvidoere Study Group on childhood d. Proinsulin, GLP-1, and glucagon are associated with partial remission in children and adolescents with newly diagnosed type 1 diabetes. Pediatr Diabetes 13(1):51–58

Karges B, Durinovic-Bello I, Heinze E, Boehm BO, Debatin KM, Karges W (2004) Complete long-term recovery of beta-cell function in autoimmune type 1 diabetes after insulin treatment. Diabetes Care 27(5):1207–1208

Karnik SK, Chen H, McLean GW, Heit JJ, Gu X, Zhang AY, Fontaine M, Yen MH, Kim SK (2007) Menin controls growth of pancreatic beta-cells in pregnant mice and promotes gestational diabetes mellitus. Science 318(5851):806–809

Kassem SA, Ariel I, Thornton PS, Scheimberg I, Glaser B (2000) Beta-cell proliferation and apoptosis in the developing normal human pancreas and in hyperinsulinism of infancy. Diabetes 49(8):1325–1333

Kasuga M (2006) Insulin resistance and pancreatic beta cell failure. J Clin Invest 116(7):1756–1760

Katsuta H, Akashi T, Katsuta R, Nagaya M, Kim D, Arinobu Y, Hara M, Bonner-Weir S, Sharma AJ, Akashi K, Weir GC (2010) Single pancreatic beta cells co-express multiple islet hormone genes in mice. Diabetologia 53(1):128–138

Katsuta H, Aguayo-Mazzucato C, Katsuta R, Akashi T, Hollister-Lock J, Sharma AJ, Bonner-Weir S, Weir GC (2012) Subpopulations of GFP-marked mouse pancreatic beta-cells differ in size, granularity, and insulin secretion. Endocrinology 153(11):5180–5187

Kiekens R, In't Veld P, Mahler T, Schuit F, Van De Winkel M, Pipeleers D (1992) Differences in glucose recognition by individual rat pancreatic B cells are associated with intercellular differences in glucose-induced biosynthetic activity. J Clin Invest 89(1):117–125

Kim K, Park SM (2018) Association of muscle mass and fat mass with insulin resistance and the prevalence of metabolic syndrome in Korean adults: a cross-sectional study. Sci Rep 8(1):2703

Kim H, Toyofuku Y, Lynn FC, Chak E, Uchida T, Mizukami H, Fujitani Y, Kawamori R, Miyatsuka T, Kosaka Y, Yang K, Honig G, van der Hart M, Kishimoto N, Wang J, Yagihashi S, Tecott LH, Watada H, German MS (2010) Serotonin regulates pancreatic beta cell mass during pregnancy. Nat Med 16(7):804–808

Kim JH, Park HY, Bae S, Lim Y-H, Hong Y-C (2013) Diethylhexyl phthalates is associated with insulin resistance via oxidative stress in the elderly: a panel study. PLoS One 8(8):e71392

Kloppel G, Drenck CR, Oberholzer M, Heitz PU (1984) Morphometric evidence for a striking B-cell reduction at the clinical onset of type 1 diabetes. Virchows Arch A Pathol Anat Histopathol 403(4):441–452

Kloppel G, Lohr M, Habich K, Oberholzer M, Heitz PU (1985) Islet pathology and the pathogenesis of type 1 and type 2 diabetes mellitus revisited. Surv Synth Pathol Res 4(2):110–125

Komatsu M, Sato Y, Aizawa T, Hashizume K (2001) KATP channel-independent glucose action: an elusive pathway in stimulus-secretion coupling of pancreatic beta-cell. Endocr J 48(3):275–288

Kopp JL, Dubois CL, Schaffer AE, Hao E, Shih HP, Seymour PA, Ma J, Sander M (2011) Sox9+ ductal cells are multipotent progenitors throughout development but do not produce new endocrine cells in the normal or injured adult pancreas. Development 138(4):653–665

Kravets V, Dwulet JM, Schleicher WE, Hodson DJ, Davis AM, Pyle L, Piscopio RA, Sticco-Ivins M, Benninger RKP (2022) Functional architecture of pancreatic islets identifies a population of first responder cells that drive the first-phase calcium response. PLoS Biol 20(9):e3001761

Kulkarni RN, Mizrachi EB, Ocana AG, Stewart AF (2012) Human beta-cell proliferation and intracellular signaling: driving in the dark without a road map. Diabetes 61(9):2205–2213

La Manno G, Soldatov R, Zeisel A, Braun E, Hochgerner H, Petukhov V, Lidschreiber K, Kastriti ME, Lönnerberg P, Furlan A, Fan J, Borm LE, Liu Z, van Bruggen D, Guo J, He X, Barker R,

Sundström E, Castelo-Branco G, Cramer P, Adameyko I, Linnarsson S, Kharchenko PV (2018) RNA velocity of single cells. Nature 560(7719):494–498

Lam CJ, Jacobson DR, Rankin MM, Cox AR, Kushner JA (2017) β cells persist in T1D pancreata without evidence of ongoing β-cell turnover or neogenesis. J Clin Endocrinol Metab 102(8):2647–2659

Lam CJ, Cox AR, Jacobson DR, Rankin MM, Kushner JA (2018) Highly proliferative alpha-cell-related islet endocrine cells in human pancreata. Diabetes 67(4):674–686

Lang DA, Matthews DR, Burnett M, Turner RC (1981) Brief, irregular oscillations of basal plasma insulin and glucose concentrations in diabetic man. Diabetes 30(5):435–439

Lawlor N, George J, Bolisetty M, Kursawe R, Sun L, Sivakamasundari V, Kycia I, Robson P, Stitzel ML (2017) Single-cell transcriptomes identify human islet cell signatures and reveal cell-type-specific expression changes in type 2 diabetes. Genome Res 27(2):208–222

Layden BT, Durai V, Newman MV, Marinelarena AM, Ahn CW, Feng G, Lin S, Zhang X, Kaufman DB, Jafari N, Sorensen GL, Lowe WL Jr (2010) Regulation of pancreatic islet gene expression in mouse islets by pregnancy. J Endocrinol 207(3):265–279

Lee Y, Hirose H, Ohneda M, Johnson JH, McGarry JD, Unger RH (1994) Beta-cell lipotoxicity in the pathogenesis of non-insulin-dependent diabetes mellitus of obese rats: impairment in adipocyte-beta-cell relationships. Proc Natl Acad Sci USA 91(23):10878–10882

Lee Y, Lingvay I, Szczepaniak LS, Ravazzola M, Orci L, Unger RH (2010) Pancreatic steatosis: harbinger of type 2 diabetes in obese rodents. Int J Obes 34(2):396–400

Lee H, Lee Y-S, Harenda Q, Pietrzak S, Oktay HZ, Schreiber S, Liao Y, Sonthalia S, Ciecko AE, Chen Y-G, Keles S, Sridharan R, Engin F (2020) Beta cell dedifferentiation induced by IRE1α deletion prevents type 1 diabetes. Cell Metab 31(4):822–836.e825

Lee S, Zhang J, Saravanakumar S, Flisher MF, Grimm DR, van der Meulen T, Huising MO (2021) Virgin beta-cells at the neogenic niche proliferate normally and mature slowly. Diabetes 70(5):1070–1083

Lee K, Chan JY, Liang C, Ip CK, Shi Y-C, Herzog H, Hughes WE, Bensellam M, Delghingaro-Augusto V, Koina ME, Nolan CJ, Laybutt DR (2022) XBP1 maintains beta cell identity, represses beta-to-alpha cell transdifferentiation and protects against diabetic beta cell failure during metabolic stress in mice. Diabetologia 65(6):984–996

Lemper M, Leuckx G, Heremans Y, German MS, Heimberg H, Bouwens L, Baeyens L (2015) Reprogramming of human pancreatic exocrine cells to beta-like cells. Cell Death Differ 22(7):1117–1130

Li J, Klughammer J, Farlik M, Penz T, Spittler A, Barbieux C, Berishvili E, Bock C, Kubicek S (2016) Single-cell transcriptomes reveal characteristic features of human pancreatic islet cell types. EMBO Rep 17(2):178–187

Li X, Yang KY, Chan VW, Leung KT, Zhang XB, Wong AS, Chong CCN, Wang CC, Ku M, Lui KO (2020) Single-cell RNA-seq reveals that CD9 is a negative marker of glucose-responsive pancreatic beta-like cells derived from human pluripotent stem cells. Stem Cell Rep 15(5):1111–1126

Li J, Zheng Y, Yan P, Song M, Wang S, Sun L, Liu Z, Ma S, Izpisua Belmonte JC, Chan P, Zhou Q, Zhang W, Liu GH, Tang F, Qu J (2021) A single-cell transcriptomic atlas of primate pancreatic islet aging. Natl Sci Rev 8(2):nwaa127

Lillioja S, Mott DM, Spraul M, Ferraro R, Foley JE, Ravussin E, Knowler WC, Bennett PH, Bogardus C (1993) Insulin resistance and insulin secretory dysfunction as precursors of non-insulin-dependent diabetes mellitus. Prospective studies of Pima Indians. N Engl J Med 329(27):1988–1992

Lin HV, Ren H, Samuel VT, Lee H-Y, Lu TY, Shulman GI, Accili D (2011) Diabetes in mice with selective impairment of insulin action in Glut4-expressing tissues. Diabetes 60(3):700–709

Linnemann AK, Baan M, Davis DB (2014) Pancreatic β-cell proliferation in obesity. Adv Nutr 5(3):278–288

Liu M, Yao C, Wang Z, Li K (2016) Polymorphism− 116C/G of the human X box binding protein 1 gene is associated with risk of type 2 diabetes in a Chinese Han population. Gene 575(1):71–74

Lu TT-H, Heyne S, Dror E, Casas E, Leonhardt L, Boenke T, Yang C-H, Sagar K, Arrigoni L, Dalgaard K, Teperino R, Enders L, Selvaraj M, Ruf M, Raja SJ, Xie H, Boenisch U, Orkin SH, Lynn FC, Hoffman BG, Grün D, Vavouri T, Lempradl AM, Pospisilik JA (2018) The polycomb-dependent epigenome controls β cell dysfunction, dedifferentiation, and diabetes. Cell Metab 27(6):1294–1308.e1297

Marchetti P, Del Guerra S, Marselli L, Lupi R, Masini M, Pollera M, Bugliani M, Boggi U, Vistoli F, Mosca F, Del Prato S (2004) Pancreatic islets from type 2 diabetic patients have functional defects and increased apoptosis that are ameliorated by metformin. J Clin Endocrinol Metab 89(11):5535–5541

Marchetti P, Bugliani M, Lupi R, Marselli L, Masini M, Boggi U, Filipponi F, Weir GC, Eizirik DL, Cnop M (2007) The endoplasmic reticulum in pancreatic beta cells of type 2 diabetes patients. Diabetologia 50(12):2486–2494

Marroqui L, Masini M, Merino B, Grieco FA, Millard I, Dubois C, Quesada I, Marchetti P, Cnop M, Eizirik DL (2015) Pancreatic alpha cells are resistant to metabolic stress-induced apoptosis in type 2 diabetes. EBioMedicine 2(5):378–385

Marselli L, Suleiman M, Masini M, Campani D, Bugliani M, Syed F, Martino L, Focosi D, Scatena F, Olimpico F, Filipponi F, Masiello P, Boggi U, Marchetti P (2014) Are we overestimating the loss of beta cells in type 2 diabetes? Diabetologia 57(2):362–365

Matsumoto I, Sawada T, Nakano M, Sakai T, Liu B, Ansite JD, Zhang HJ, Kandaswamy R, Sutherland DE, Hering BJ (2004) Improvement in islet yield from obese donors for human islet transplants. Transplantation 78(6):880–885

Meier JJ, Bhushan A, Butler AE, Rizza RA, Butler PC (2005) Sustained beta cell apoptosis in patients with long-standing type 1 diabetes: indirect evidence for islet regeneration? Diabetologia 48(11):2221–2228

Meier JJ, Lin JC, Butler AE, Galasso R, Martinez DS, Butler PC (2006) Direct evidence of attempted beta cell regeneration in an 89-year-old patient with recent-onset type 1 diabetes. Diabetologia 49(8):1838–1844

Meier JJ, Butler AE, Saisho Y, Monchamp T, Galasso R, Bhushan A, Rizza RA, Butler PC (2008) Beta-cell replication is the primary mechanism subserving the postnatal expansion of beta-cell mass in humans. Diabetes 57(6):1584–1594

Meier JJ, Köhler CU, Alkhatib B, Sergi C, Junker T, Klein HH, Schmidt WE, Fritsch H (2010) Beta-cell development and turnover during prenatal life in humans. Eur J Endocrinol 162(3):559–568

Menge BA, Tannapfel A, Belyaev O, Drescher R, Muller C, Uhl W, Schmidt WE, Meier JJ (2008) Partial pancreatectomy in adult humans does not provoke beta-cell regeneration. Diabetes 57(1):142–149

Meter D, Racetin A, Vukojević K, Balog M, Ivić V, Zjalić M, Heffer M, Filipović N (2022) A lack of GD3 synthase leads to impaired renal expression of connexins and Pannexin1 in St8sia1 knockout mice. Int J Mol Sci 23(11):6237

Mezghenna K, Pomies P, Chalancon A, Castex F, Leroy J, Niclauss N, Nadal B, Cambier L, Cazevieille C, Petit P, Gomis R, Berney T, Gross R, Lajoix AD (2011) Increased neuronal nitric oxide synthase dimerisation is involved in rat and human pancreatic beta cell hyperactivity in obesity. Diabetologia 54(11):2856–2866

Mezza T, Muscogiuri G, Sorice GP, Clemente G, Hu J, Pontecorvi A, Holst JJ, Giaccari A, Kulkarni RN (2014) Insulin resistance alters islet morphology in nondiabetic humans. Diabetes 63(3):994–1007

Mills JC, Stanger BZ, Sander M (2019) Nomenclature for cellular plasticity: are the terms as plastic as the cells themselves? EMBO J 38(19):e103148

Minami K, Okuno M, Miyawaki K, Okumachi A, Ishizaki K, Oyama K, Kawaguchi M, Ishizuka N, Iwanaga T, Seino S (2005) Lineage tracing and characterization of insulin-secreting cells generated from adult pancreatic acinar cells. Proc Natl Acad Sci USA 102(42):15116–15121

Mosser RE, Maulis MF, Moulle VS, Dunn JC, Carboneau BA, Arasi K, Pappan K, Poitout V, Gannon M (2015) High-fat diet-induced beta-cell proliferation occurs prior to insulin resistance in C57Bl/6J male mice. Am J Physiol Endocrinol Metab 308(7):E573–E582

Muraro MJ, Dharmadhikari G, Grun D, Groen N, Dielen T, Jansen E, van Gurp L, Engelse MA, Carlotti F, de Koning EJ, van Oudenaarden A (2016) A single-cell transcriptome atlas of the human pancreas. Cell Syst 3(4):385–394 e383

Nesher R, Cerasi E (2002) Modeling phasic insulin release: immediate and time-dependent effects of glucose. Diabetes 51(Suppl 1):S53–S59

Nir T, Melton DA, Dor Y (2007) Recovery from diabetes in mice by beta cell regeneration. J Clin Invest 117(9):2553–2561

Nunemaker CS, Wasserman DH, McGuinness OP, Sweet IR, Teague JC, Satin LS (2006) Insulin secretion in the conscious mouse is biphasic and pulsatile. Am J Physiol Endocrinol Metab 290(3):E523–E529

Oram RA, Sims EK, Evans-Molina C (2019) Beta cells in type 1 diabetes: mass and function; sleeping or dead? Diabetologia 62(4):567–577

Orino K, Lehman L, Tsuji Y, Ayaki H, Torti SV, Torti FM (2001) Ferritin and the response to oxidative stress. Biochem J 357(Pt 1):241–247

Osei K, Gaillard T, Schuster DP (1997) Pathogenetic mechanisms of impaired glucose tolerance and type II diabetes in African-Americans. The significance of insulin secretion, insulin sensitivity, and glucose effectiveness. Diabetes Care 20(3):396–404

Pagliuca FW, Millman JR, Gürtler M, Segel M, Van Dervort A, Ryu JH, Peterson QP, Greiner D, Melton DA (2014) Generation of functional human pancreatic β cells in vitro. Cell 159(2):428–439

Pan FC, Bankaitis ED, Boyer D, Xu X, Van de Casteele M, Magnuson MA, Heimberg H, Wright CVE (2013) Spatiotemporal patterns of multipotentiality in Ptf1a-expressing cells during pancreas organogenesis and injury-induced facultative restoration. Development 140(4):751–764

Parsons JA, Brelje TC, Sorenson RL (1992) Adaptation of islets of Langerhans to pregnancy: increased islet cell proliferation and insulin secretion correlates with the onset of placental lactogen secretion. Endocrinology 130(3):1459–1466

Parsons JA, Bartke A, Sorenson RL (1995) Number and size of islets of Langerhans in pregnant, human growth hormone-expressing transgenic, and pituitary dwarf mice: effect of lactogenic hormones. Endocrinology 136(5):2013–2021

Perelis M, Marcheva B, Ramsey KM, Schipma MJ, Hutchison AL, Taguchi A, Peek CB, Hong H, Huang W, Omura C, Allred AL, Bradfield CA, Dinner AR, Barish GD, Bass J (2015) Pancreatic β cell enhancers regulate rhythmic transcription of genes controlling insulin secretion. Science 350(6261):aac4250

Perl S, Kushner JA, Buchholz BA, Meeker AK, Stein GM, Hsieh M, Kirby M, Pechhold S, Liu EH, Harlan DM, Tisdale JF (2010) Significant human β-cell turnover is limited to the first three decades of life as determined by in vivo thymidine analog incorporation and radiocarbon dating. J Clin Endocrinol Metab 95(10):E234–E239

Perley M, Kipnis DM (1966) Plasma insulin responses to glucose and tolbutamide of normal weight and obese diabetic and nondiabetic subjects. Diabetes 15(12):867–874

Pick A, Clark J, Kubstrup C, Levisetti M, Pugh W, Bonner-Weir S, Polonsky KS (1998) Role of apoptosis in failure of beta-cell mass compensation for insulin resistance and beta-cell defects in the male Zucker diabetic fatty rat. Diabetes 47(3):358–364

Piran R, Lee SH, Li CR, Charbono A, Bradley LM, Levine F (2014) Pharmacological induction of pancreatic islet cell transdifferentiation: relevance to type I diabetes. Cell Death Dis 5:e1357

Piston DW, Knobel SM, Postic C, Shelton KD, Magnuson MA (1999) Adenovirus-mediated knockout of a conditional glucokinase gene in isolated pancreatic islets reveals an essential role for proximal metabolic coupling events in glucose-stimulated insulin secretion. J Biol Chem 274(2):1000–1004

Polonsky KS, Given BD, Van Cauter E (1988a) Twenty-four-hour profiles and pulsatile patterns of insulin secretion in normal and obese subjects. J Clin Invest 81(2):442–448

Polonsky KS, Given BD, Hirsch LJ, Tillil H, Shapiro ET, Beebe C, Frank BH, Galloway JA, Van Cauter E (1988b) Abnormal patterns of insulin secretion in non-insulin-dependent diabetes mellitus. N Engl J Med 318(19):1231–1239

Porksen N (2002) The in vivo regulation of pulsatile insulin secretion. Diabetologia 45(1):3–20
Porte D Jr, Pupo AA (1969) Insulin responses to glucose: evidence for a two pool system in man. J Clin Invest 48(12):2309–2319
Rahier J, Guiot Y, Goebbels RM, Sempoux C, Henquin JC (2008) Pancreatic beta-cell mass in European subjects with type 2 diabetes. Diabetes Obes Metab 10(Suppl 4):32–42
Rakshit K, Qian J, Ernst J, Matveyenko AV (2016) Circadian variation of the pancreatic islet transcriptome. Physiol Genomics 48(9):677–687
Rankin MM, Kushner JA (2009) Adaptive beta-cell proliferation is severely restricted with advanced age. Diabetes 58(6):1365–1372
Rantanen K, Pursiheimo JP, Hogel H, Miikkulainen P, Sundstrom J, Jaakkola PM (2013) p62/SQSTM1 regulates cellular oxygen sensing by attenuating PHD3 activity through aggregate sequestration and enhanced degradation. J Cell Sci 126(Pt 5):1144–1154
Rieck S, Kaestner KH (2010) Expansion of beta-cell mass in response to pregnancy. Trends Endocrinol Metab 21(3):151–158
Rieck S, White P, Schug J, Fox AJ, Smirnova O, Gao N, Gupta RK, Wang ZV, Scherer PE, Keller MP, Attie AD, Kaestner KH (2009) The transcriptional response of the islet to pregnancy in mice. Mol Endocrinol 23(10):1702–1712
Rodriguez-Calvo T, Zapardiel-Gonzalo J, Amirian N, Castillo E, Lajevardi Y, Krogvold L, Dahl-Jorgensen K, von Herrath MG (2017) Increase in pancreatic proinsulin and preservation of beta-cell mass in autoantibody-positive donors prior to type 1 diabetes onset. Diabetes 66(5):1334–1345
Rorsman P, Renström E (2003) Insulin granule dynamics in pancreatic beta cells. Diabetologia 46(8):1029–1045
Rui J, Deng S, Arazi A, Perdigoto AL, Liu Z, Herold KC (2017) Beta cells that resist immunological attack develop during progression of autoimmune diabetes in NOD mice. Cell Metab 25(3):727–738
Sahin GS, Lee H, Engin F (2021) An accomplice more than a mere victim: the impact of beta-cell ER stress on type 1 diabetes pathogenesis. Mol Metab 54:101365
Saisho Y, Butler AE, Manesso E, Elashoff D, Rizza RA, Butler PC (2013) Beta-cell mass and turnover in humans: effects of obesity and aging. Diabetes Care 36(1):111–117
Sakuraba H, Mizukami H, Yagihashi N, Wada R, Hanyu C, Yagihashi S (2002) Reduced beta-cell mass and expression of oxidative stress-related DNA damage in the islet of Japanese Type II diabetic patients. Diabetologia 45(1):85–96
Salem V, Silva LD, Suba K, Georgiadou E, Neda Mousavy Gharavy S, Akhtar N, Martin-Alonso A, Gaboriau DCA, Rothery SM, Stylianides T, Carrat G, Pullen TJ, Singh SP, Hodson DJ, Leclerc I, Shapiro AMJ, Marchetti P, Briant LJB, Distaso W, Ninov N, Rutter GA (2019) Leader beta-cells coordinate Ca(2+) dynamics across pancreatic islets in vivo. Nat Metab 1(6):615–629
Salomon D, Meda P (1986) Heterogeneity and contact-dependent regulation of hormone secretion by individual B cells. Exp Cell Res 162(2):507–520
Scaglia L, Smith FE, Bonner-Weir S (1995) Apoptosis contributes to the involution of beta cell mass in the post partum rat pancreas. Endocrinology 136(12):5461–5468
Scarl RT, Corbin KL, Vann NW, Smith HM, Satin LS, Sherman A, Nunemaker CS (2019) Intact pancreatic islets and dispersed beta-cells both generate intracellular calcium oscillations but differ in their responsiveness to glucose. Cell Calcium 83:102081
Scholin A, Nystrom L, Arnqvist H, Bolinder J, Bjork E, Berne C, Karlsson FA (2011) Diabetes Incidence Study Group in S. Proinsulin/C-peptide ratio, glucagon and remission in new-onset type 1 diabetes mellitus in young adults. Diabet Med 28(2):156–161
Schroder M, Kaufman RJ (2005) ER stress and the unfolded protein response. Mutat Res 569(1–2):29–63
Segerstolpe A, Palasantza A, Eliasson P, Andersson EM, Andreasson AC, Sun X, Picelli S, Sabirsh A, Clausen M, Bjursell MK, Smith DM, Kasper M, Ammala C, Sandberg R (2016) Single-cell transcriptome profiling of human pancreatic islets in health and type 2 diabetes. Cell Metab 24(4):593–607

Seltzer HS, Allen EW, Herron AL Jr, Brennan MT (1967) Insulin secretion in response to glycemic stimulus: relation of delayed initial release to carbohydrate intolerance in mild diabetes mellitus. J Clin Invest 46(3):323–335

Seshadri N, Doucette CA (2021) Circadian regulation of the pancreatic beta cell. Endocrinology 162(9):bqab089

Sharma RB, O'Donnell AC, Stamateris RE, Ha B, McCloskey KM, Reynolds PR, Arvan P, Alonso LC (2015) Insulin demand regulates beta cell number via the unfolded protein response. J Clin Invest 125(10):3831–3846

Sharma RB, Landa-Galván HV, Alonso LC (2021) Living dangerously: protective and harmful ER stress responses in pancreatic β-cells. Diabetes 70(11):2431–2443

Sherry NA, Kushner JA, Glandt M, Kitamura T, Brillantes AM, Herold KC (2006) Effects of autoimmunity and immune therapy on beta-cell turnover in type 1 diabetes. Diabetes 55(12):3238–3245

Singh SP, Janjuha S, Hartmann T, Kayisoglu O, Konantz J, Birke S, Murawala P, Alfar EA, Murata K, Eugster A, Tsuji N, Morrissey ER, Brand M, Ninov N (2017) Different developmental histories of beta-cells generate functional and proliferative heterogeneity during islet growth. Nat Commun 8(1):664

Solar M, Cardalda C, Houbracken I, Martin M, Maestro MA, De Medts N, Xu X, Grau V, Heimberg H, Bouwens L, Ferrer J (2009) Pancreatic exocrine duct cells give rise to insulin-producing beta cells during embryogenesis but not after birth. Dev Cell 17(6):849–860

Solas M, Aisa B, Tordera RM, Mugueta MC, Ramírez MJ (2013) Stress contributes to the development of central insulin resistance during aging: implications for Alzheimer's disease. Biochim Biophys Acta 1832(12):2332–2339

Sorenson RL, Brelje TC (1997) Adaptation of islets of Langerhans to pregnancy: beta-cell growth, enhanced insulin secretion and the role of lactogenic hormones. Horm Metab Res 29(6):301–307

Spijker HS, Song H, Ellenbroek JH, Roefs MM, Engelse MA, Bos E, Koster AJ, Rabelink TJ, Hansen BC, Clark A, Carlotti F, de Koning EJP (2015) Loss of β-cell identity occurs in type 2 diabetes and is associated with islet amyloid deposits. Diabetes 64(8):2928–2938

Sreenan S, Pick AJ, Levisetti M, Baldwin AC, Pugh W, Polonsky KS (1999) Increased beta-cell proliferation and reduced mass before diabetes onset in the nonobese diabetic mouse. Diabetes 48(5):989–996

Stefan Y, Meda P, Neufeld M, Orci L (1987) Stimulation of insulin secretion reveals heterogeneity of pancreatic B cells in vivo. J Clin Invest 80(1):175–183

Straub SG, Sharp GWG (2002) Glucose-stimulated signaling pathways in biphasic insulin secretion. Diabetes Metab Res Rev 18(6):451–463

Straub SG, Sharp GWG (2004) Hypothesis: one rate-limiting step controls the magnitude of both phases of glucose-stimulated insulin secretion. Am J Physiol Cell Physiol 287(3):C565–C571

Sun J, Ni Q, Xie J, Xu M, Zhang J, Kuang J, Wang Y, Ning G, Wang Q (2019) β-cell dedifferentiation in patients with T2D with adequate glucose control and nondiabetic chronic pancreatitis. J Clin Endocrinol Metab 104(1):83–94

Szabat M, Pourghaderi P, Soukhatcheva G, Verchere CB, Warnock GL, Piret JM, Johnson JD (2011) Kinetics and genomic profiling of adult human and mouse beta-cell maturation. Islets 3(4):175–187

Talchai C, Xuan S, Lin HV, Sussel L, Accili D (2012) Pancreatic beta cell dedifferentiation as a mechanism of diabetic beta cell failure. Cell 150(6):1223–1234

Tang R, Zhong T, Wu C, Zhou Z, Li X (2019) The remission phase in type 1 diabetes: role of hyperglycemia rectification in immune modulation. Front Endocrinol 10:824

Tatsuoka H, Sakamoto S, Yabe D, Kabai R, Kato U, Okumura T, Botagarova A, Tokumoto S, Usui R, Ogura M, Nagashima K, Mukai E, Fujitani Y, Watanabe A, Inagaki N (2020) Single-cell transcriptome analysis dissects the replicating process of pancreatic beta cells in partial pancreatectomy model. iScience 23(12):101774

Thorel F, Nepote V, Avril I, Kohno K, Desgraz R, Chera S, Herrera PL (2010) Conversion of adult pancreatic alpha-cells to beta-cells after extreme beta-cell loss. Nature 464(7292):1149–1154
Tosti L, Hang Y, Debnath O, Tiesmeyer S, Trefzer T, Steiger K, Ten FW, Lukassen S, Ballke S, Kühl AA, Spieckermann S, Bottino R, Ishaque N, Weichert W, Kim SK, Eils R, Conrad C (2021) Single-nucleus and in situ RNA–sequencing reveal cell topographies in the human pancreas. Gastroenterology 160(4):1330–1344.e1311
Trapnell C, Cacchiarelli D, Grimsby J, Pokharel P, Li S, Morse M, Lennon NJ, Livak KJ, Mikkelsen TS, Rinn JL (2014) The dynamics and regulators of cell fate decisions are revealed by pseudotemporal ordering of single cells. Nat Biotechnol 32(4):381–386
Tschen SI, Dhawan S, Gurlo T, Bhushan A (2009) Age-dependent decline in beta-cell proliferation restricts the capacity of beta-cell regeneration in mice. Diabetes 58(6):1312–1320
Van Assche FA, Aerts L, De Prins F (1978) A morphological study of the endocrine pancreas in human pregnancy. Br J Obstet Gynaecol 85(11):818–820
Van De Winkel M, Pipeleers D (1983) Autofluorescence-activated cell sorting of pancreatic islet cells: purification of insulin-containing B-cells according to glucose-induced changes in cellular redox state. Biochem Biophys Res Commun 114(2):835–842
van der Meulen T, Mawla AM, DiGruccio MR, Adams MW, Nies V, Dolleman S, Liu S, Ackermann AM, Caceres E, Hunter AE, Kaestner KH, Donaldson CJ, Huising MO (2017) Virgin beta cells persist throughout life at a neogenic niche within pancreatic islets. Cell Metab 25(4):911–926 e916
Van Schravendijk CF, Kiekens R, Pipeleers DG (1992) Pancreatic beta cell heterogeneity in glucose-induced insulin secretion. J Biol Chem 267(30):21344–21348
Vasavada RC, Garcia-Ocana A, Zawalich WS, Sorenson RL, Dann P, Syed M, Ogren L, Talamantes F, Stewart AF (2000) Targeted expression of placental lactogen in the beta cells of transgenic mice results in beta cell proliferation, islet mass augmentation, and hypoglycemia. J Biol Chem 275(20):15399–15406
Wang YJ, Kaestner KH (2019) Single-cell RNA-seq of the pancreatic islets—a promise not yet fulfilled? Cell Metab 29(3):539–544
Wang W, Zhang C (2021) Targeting β-cell dedifferentiation and transdifferentiation: opportunities and challenges. Endocr Connect 10(8):R213–R228
Wang RN, Kloppel G, Bouwens L (1995) Duct- to islet-cell differentiation and islet growth in the pancreas of duct-ligated adult rats. Diabetologia 38(12):1405–1411
Wang YJ, Schug J, Won KJ, Liu C, Naji A, Avrahami D, Golson ML, Kaestner KH (2016a) Single-cell transcriptomics of the human endocrine pancreas. Diabetes 65(10):3028–3038
Wang YJ, Golson ML, Schug J, Traum D, Liu C, Vivek K, Dorrell C, Naji A, Powers AC, Chang KM, Grompe M, Kaestner KH (2016b) Single-cell mass cytometry analysis of the human endocrine pancreas. Cell Metab 24(4):616–626
Wang YJ, Traum D, Schug J, Gao L, Liu C, Consortium H, Atkinson MA, Powers AC, Feldman MD, Naji A, Chang K-M, Kaestner KH (2019) Multiplexed in situ imaging mass cytometry analysis of the human endocrine pancreas and immune system in type 1 diabetes. Cell Metab 29(3):769–783.e764
Ward WK, Bolgiano DC, McKnight B, Halter JB, Porte D Jr (1984) Diminished B cell secretory capacity in patients with noninsulin-dependent diabetes mellitus. J Clin Invest 74(4):1318–1328
Westacott MJ, Ludin NWF, Benninger RKP (2017) Spatially organized β-cell subpopulations control electrical dynamics across islets of langerhans. Biophys J 113(5):1093–1108
Willcox A, Richardson SJ, Bone AJ, Foulis AK, Morgan NG (2010) Evidence of increased islet cell proliferation in patients with recent-onset type 1 diabetes. Diabetologia 53(9):2020–2028
Wojtusciszyn A, Armanet M, Morel P, Berney T, Bosco D (2008) Insulin secretion from human beta cells is heterogeneous and dependent on cell-to-cell contacts. Diabetologia 51(10):1843–1852
Wollny D, Zhao S, Everlien I, Lun X, Brunken J, Brune D, Ziebell F, Tabansky I, Weichert W, Marciniak-Czochra A, Martin-Villalba A (2016) Single-cell analysis uncovers clonal acinar cell heterogeneity in the adult pancreas. Dev Cell 39(3):289–301

Wood AR, Jonsson A, Jackson AU, Wang N, van Leewen N, Palmer ND, Kobes S, Deelen J, Boquete-Vilarino L, Paananen J, Stančáková A, Boomsma DI, de Geus EJC, Eekhoff EMW, Fritsche A, Kramer M, Nijpels G, Simonis-Bik A, van Haeften TW, Mahajan A, Boehnke M, Bergman RN, Tuomilehto J, Collins FS, Mohlke KL, Banasik K, Groves CJ, McCarthy MI, Diabetes Research on Patient S, Pearson ER, Natali A, Mari A, Buchanan TA, Taylor KD, Xiang AH, Gjesing AP, Grarup N, Eiberg H, Pedersen O, Chen Y-D, Laakso M, Norris JM, Smith U, Wagenknecht LE, Baier L, Bowden DW, Hansen T, Walker M, Watanabe RM, Hart LM, Hanson RL, Frayling TM (2017) A genome-wide association study of IVGTT-based measures of first-phase insulin secretion refines the underlying physiology of type 2 diabetes variants. Diabetes 66(8):2296–2309

Xin Y, Kim J, Okamoto H, Ni M, Wei Y, Adler C, Murphy AJ, Yancopoulos GD, Lin C, Gromada J (2016) RNA sequencing of single human islet cells reveals type 2 diabetes genes. Cell Metab 24(4):608–615

Xin Y, Dominguez Gutierrez G, Okamoto H, Kim J, Lee AH, Adler C, Ni M, Yancopoulos GD, Murphy AJ, Gromada J (2018) Pseudotime ordering of single human beta-cells reveals states of insulin production and unfolded protein response. Diabetes 67(9):1783–1794

Xu X, D'Hoker J, Stange G, Bonne S, De Leu N, Xiao X, Van de Casteele M, Mellitzer G, Ling Z, Pipeleers D, Bouwens L, Scharfmann R, Gradwohl G, Heimberg H (2008) Beta cells can be generated from endogenous progenitors in injured adult mouse pancreas. Cell 132(2):197–207

Yamada T, Cavelti-Weder C, Caballero F, Lysy PA, Guo L, Sharma A, Li W, Zhou Q, Bonner-Weir S, Weir GC (2015) Reprogramming mouse cells with a pancreatic duct phenotype to insulin-producing beta-like cells. Endocrinology 156(6):2029–2038

Ye L, Robertson MA, Hesselson D, Stainier DY, Anderson RM (2015) Glucagon is essential for alpha cell transdifferentiation and beta cell neogenesis. Development 142(8):1407–1417

Yoneda S, Uno S, Iwahashi H, Fujita Y, Yoshikawa A, Kozawa J, Okita K, Takiuchi D, Eguchi H, Nagano H, Imagawa A, Shimomura I (2013) Predominance of beta-cell neogenesis rather than replication in humans with an impaired glucose tolerance and newly diagnosed diabetes. J Clin Endocrinol Metab 98(5):2053–2061

Yong HJ, Xie G, Liu C, Wang W, Naji A, Irianto J, Wang YJ (2021) Gene signatures of NEUROGENIN3+ endocrine progenitor cells in the human pancreas. Front Endocrinol 12:736286

Yong HJ, Toledo MP, Nowakowski RS, Wang YJ (2022) Sex differences in the molecular programs of pancreatic cells contribute to the differential risks of type 2 diabetes. Endocrinology 163(11):bqac156

Yoon KH, Ko SH, Cho JH, Lee JM, Ahn YB, Song KH, Yoo SJ, Kang MI, Cha BY, Lee KW, Son HY, Kang SK, Kim HS, Lee IK, Bonner-Weir S (2003) Selective beta-cell loss and alpha-cell expansion in patients with type 2 diabetes mellitus in Korea. J Clin Endocrinol Metab 88(5):2300–2308

Yu X-X, Qiu W-L, Yang L, Wang Y-C, He M-Y, Wang D, Zhang Y, Li L-C, Zhang J, Wang Y, Xu C-R (2021) Sequential progenitor states mark the generation of pancreatic endocrine lineages in mice and humans. Cell Res 31(8):886–903

Zhang H, Zhang J, Pope CF, Crawford LA, Vasavada RC, Jagasia SM, Gannon M (2010) Gestational diabetes mellitus resulting from impaired beta-cell compensation in the absence of FoxM1, a novel downstream effector of placental lactogen. Diabetes 59(1):143–152

Zhao H, Huang X, Liu Z, Pu W, Lv Z, He L, Li Y, Zhou Q, Lui KO, Zhou B (2021) Pre-existing beta cells but not progenitors contribute to new beta cells in the adult pancreas. Nat Metab 3(3):352–365

Zhou Q, Brown J, Kanarek A, Rajagopal J, Melton DA (2008) In vivo reprogramming of adult pancreatic exocrine cells to beta-cells. Nature 455(7213):627–632

Zhou Y, Duan S, Zhou Y, Yu S, Wu J, Wu X, Zhao J, Zhao Y (2015) Sulfiredoxin-1 attenuates oxidative stress via Nrf2/ARE pathway and 2-Cys Prdxs after oxygen-glucose deprivation in astrocytes. J Mol Neurosci 55(4):941–950

β-Cell Regeneration Is Driven by Pancreatic Plasticity

Adrián Holguín-Horcajo, Rocio Sancho, and Meritxell Rovira

Abstract The pancreas has been considered a non-regenerative organ. β cells lost in diabetes are not replaced due to the inability of the pancreas to regenerate. However, ample evidence generated in the last few decades using murine models has demonstrated that the pancreas has a remarkable plasticity wherein differentiated cells can change cell fate toward a β-like cell phenotype. Although this process is observed after using rather artificial stimuli and the conversion efficiency is very limited, these findings have shed some light on novel pathways for β-cell regeneration. In this chapter, we will summarize the different cellular interconversion processes described to date, the experimental details and molecular regulation of such interconversions, and the genomic technologies that have allowed the identification of potential new ways to generate β cells.

Keywords Pancreas plasticity · β-cell regeneration · Diabetes · β-cell replication · Pancreas progenitors

---

A. Holguín-Horcajo · M. Rovira (✉)
Department of Physiological Science, School of Medicine, University of Barcelona, L'Hospitalet de Llobregat, Barcelona, Spain

Pancreas Regeneration: Pancreatic Progenitors and Their Niche Group, Regenerative Medicine Program, Institut d'Investigació Biomèdica de Bellvitge-IDIBELL, L'Hospitalet de Llobregat, Spain

Program for advancing the Clinical Translation of Regenerative Medicine of Catalonia, P-CMR[C], L'Hospitalet de Llobregat, Spain
e-mail: aholguin@idibell.cat; mrovira@idibell.cat

R. Sancho (✉)
Centre for Gene therapy and Regenerative Medicine, Faculty of Life Sciences & Medicine, King's College London, London, UK

Department of Medicine III, University Hospital Carl Gustav Carus, Dresden, Germany
e-mail: rocio.sancho@kcl.ac.uk

M. L. Golson (ed.), *Molecular and Cell Biology of Pancreas Development, Function and Regeneration*, Advances in Anatomy, Embryology and Cell Biology 239, https://doi.org/10.1007/978-3-031-62232-8_4

Abbreviations

ALDH1a3	Aldehyde Dehydrogenase 1 Family Member A3
AP	Alkaline phosphatase
ARX	Aristaless-related homeobox
BMPR1A	Bone Morphogenetic Protein Receptor Type 1A
BrdU	Bromodeoxyuridine
CAC	Centroacinar cell
CD24	Cluster of differentiation 24
CD51	Cluster of differentiation 51
DNMT1	DNA (cytosine-5)-methyltransferase 1
DTR	Diphtheria toxin receptor
ERK	Extracellular regulated kinase
FBW7	F-box/WD repeat-containing protein 7
FLTP1	Flattop
FOXO1	Forkhead box protein O1
GABA	Gamma-aminobutyric acid
HNF1β	Hepatocyte nuclear factor-1-β
IdU	Iododeoxyuridine
IRS	Insulin receptor substrate
IRX2	Iroquois Homeobox 2
KO	Knock out
KRT19	Keratin 19
MAFA	MAF BZIP Transcription Factor A
MAFB	MAF BZIP Transcription Factor B
mTOR	Mammalian target of rapamycin
NFAT	Nuclear factor of activated T cells
NGN3	Neurogenin 3
NKX2.2	NK2 Homeobox 2
NKX6.1	NK6 Homeobox 1
PAX4	Paired box 4
PDX1	Pancreatic and duodenal homeobox 1
PI3K	Phosphoinositide 3-kinase
PPY	Pancreatic polypeptide Y
Procr	Protein C receptor
PTF1a	Pancreas-associated transcription factor 1a
RAF	Rapidly accelerated fibrosarcoma
RAS	Rat sarcoma virus
RIP	Rat insulin promoter
rtTA	Reverse tetracycline transactivator
SOX9	SRY-Box Transcription Factor 9
SST	Somatostatin
STZ	Streptozotocin
T1D	Type 1 diabetes

T2D	Type 2 diabetes
TDC	Terminal ducta cell
TetO	Tetracyclin On
TGFβ	Transforming growth factor β
UCN3	Urocortin 3
XBP1	X-box binding protein 1

1 Introduction

The pancreas is a rather distinctive gland formed by two morphologically and functionally discrete parts: the exocrine and endocrine compartments. The exocrine pancreas is composed of acinar and ductal cells that fulfill digestive functions. The endocrine pancreas is formed by hormone-producing cells that regulate glucose homeostasis. Most of the pancreas comprises the acinar parenchyma, where a ductal network is embedded (Pan and Wright 2011; Edlund 2002). At the functional level, acinar cells produce and secrete digestive enzymes into the ducts, such as lipases, proteases, and amylases to digest lipids, proteins, and carbohydrates in the duodenum. The ductal network drives the acinar secretions into the duodenum and produces and secretes bicarbonate. The function of bicarbonate secretion by ductal cells is to neutralize not only the acid chyme entering the duodenum from the stomach, but also the acidic content secreted by acinar cells.

The other major component of the pancreas, embedded in the acinar parenchyma, is formed by the islets of Langherhans containing hormone-producing cells. Several cell types co-exist in the islet of Langerhans: β cells producing insulin, α cells producing glucagon, δ cells producing somatostatin, PP cells secreting pancreatic polypeptide, and ε cells secreting ghrelin. Of those, the most abundant are the insulin-producing β cells, accounting for between 55% and 70% of the islet cells (Pan and Wright 2011; Edlund 2002; Bonner-Weir et al. 2015). Though insulin is the primary glucose-regulating hormone, all pancreatic hormones play a role in regulating glucose homeostasis. However, the insulin-producing β cells are critical, as they are irreversibly lost in type 1 (T1D) and late stages of type 2 diabetes (T2D), leading to the inability of diabetes patients to regulate glucose homeostasis (Tan et al. 2019).

Both exocrine (acinar and ductal) and endocrine (α, β, δ, ε, and PP) cells are derived from a common progenitor during embryonic development. After posterior foregut specification, the expression of the transcription factor PDX1 induces the pancreas progenitor fate (Pan and Wright 2011). The function of PDX1 is essential for the differentiation of all pancreatic cells; pancreas agenesis occurs in *Pdx1*$^{-/-}$ mice and PDX1 loss-of-function mutations in humans (Stoffers et al. 1997; Jonsson et al. 1994). Upon the orchestration of a complex transcriptional network, PDX1^{+} progenitors give rise to endocrine cells, acinar cells, and ductal cells. A key factor involved in endocrine cell differentiation is Neurogenin 3 (NGN3), as demonstrated by *Ngn3*$^{-/-}$ mice, which are devoid of any endocrine cells (Gradwohl et al. 2000). Equally important for the differentiation of acinar and ductal cells are PTF1a and SOX9/HNF1β, respectively. Though NGN3, PTF1a, and SOX9/HNF1β are crucial for the differentiation of the three main cell types in the pancreas, the transcriptional regulation of cell fate decisions is far more complex, with a multitude of factors regulating pancreatic cell differentiation during embryonic development (Pan and Wright 2011; Edlund 2002; Bastidas-Ponce et al. 2017), which is beyond the scope of this chapter.

Diabetes mellitus is a complex multifactorial disorder. The pathogenesis of T1D and T2D is different; T1D is caused by the autoimmune destruction of β cells, while T2D seems to be a collection of diseases that result in a functionally deficient group of β cells. Some types of T2D are initially driven by insulin resistance (Kharroubi and Darwish 2015; Ahlqvist et al. 2020), while others result from a primary reduction of insulin secretion (Ahlqvist et al. 2020). In either case, restoring β-cell function is the overarching goal for improved therapy for both T1D and T2D. This goal could be achieved by cell-replacement treatment with exogenous cells or endogenous progenitors, by expanding them in vitro, or by triggering intrinsic regenerative mechanisms of the pancreas. For T1D, a combination of β-cell replacement and immunosuppressive therapy to ameliorate autoimmunity could be a curative treatment, whereas for T2D, enhancing endogenous mechanisms of β-cell regeneration and function and reducing insulin resistance might optimize blood glucose control.

In the last decade, great advances have been made in the field of β-cell replacement therapies from exogenous cells. In 2006, d'Amour et al. published the first proof-of-principle evidence that human embryonic stem cells could be differentiated into insulin-producing cells (D'Amour et al. 2006). To date, several new protocols able to efficiently differentiate mature β-like cells have been developed, leading to two phase I–II clinical trials. The results from the first trial, conducted by Viacyte, have been released recently (Ramzy et al. 2021; Shapiro et al. 2021). Although not sufficient to normalize glycemia in diabetic patients, transplanted cells survive and can sense glucose and release insulin. However, the number and maturity of cells generated, the graft rejection, and side effects due to immunosuppression call for improvements and further investigation before new clinical trials. The second clinical trial, performed by Vertex (Clinical trial NCT04786262), is based on a hepatic portal vein infusion of VX-880 cells in combination with a standard regimen of immunosuppressive agents. However, it is still in the recruitment phase; results are not yet available.

Without minimizing the importance of the abovementioned β-cell replacement therapies from exogenous cells, this chapter focuses on an alternative possibility: to regenerate β cells from endogenous pancreatic cells. Adult progenitors have been intensely studied in highly proliferative and regenerative tissues, such as the intestine and skin (McCarthy et al. 2020; Fuchs 2009). These tissues with high regenerative capacities harbor a significant pool of progenitors that comprise 5–8% of the tissue, allowing tissue renewal in homeostatic conditions or upon injury (Gonzales and Fuchs 2017; Guiu et al. 2019). The adult pancreas exhibits low physiological turnover; however, it displays some exocrine regenerative ability upon damage.

Endocrine regeneration, however, is extremely low; thus, the reservoir of progenitors is likely modest if it exists. To date, most lineage-tracing studies suggest that dedicated progenitors do not exist in this gland. However, experimental murine models using rather artificial triggers have revealed a certain degree of proliferative capacity and plasticity in the pancreas resulting from β-cell replication, intra-islet transdifferentiation, and intra-pancreatic plasticity (Demcollari et al. 2017). Furthermore, new technologies such as single-cell RNAseq (scRNAseq) have uncovered a remarkable functional heterogeneity within all the pancreatic cellular compartments (Muraro et al. 2016; Wang and Kaestner 2019; Hendley et al. 2021; Tosti et al. 2020). This chapter briefly summarizes recent efforts to allow β-cell regeneration by describing current approaches such as (Pan and Wright 2011) increasing β-cell self-replication, (Edlund 2002) conversion of other pancreatic endocrine cells into β cells, and to a lesser extent (Bonner-Weir et al. 2015) neogenesis from ductal cells and (Tan et al. 2019) conversion of acinar cells into β cells (Fig. 1).

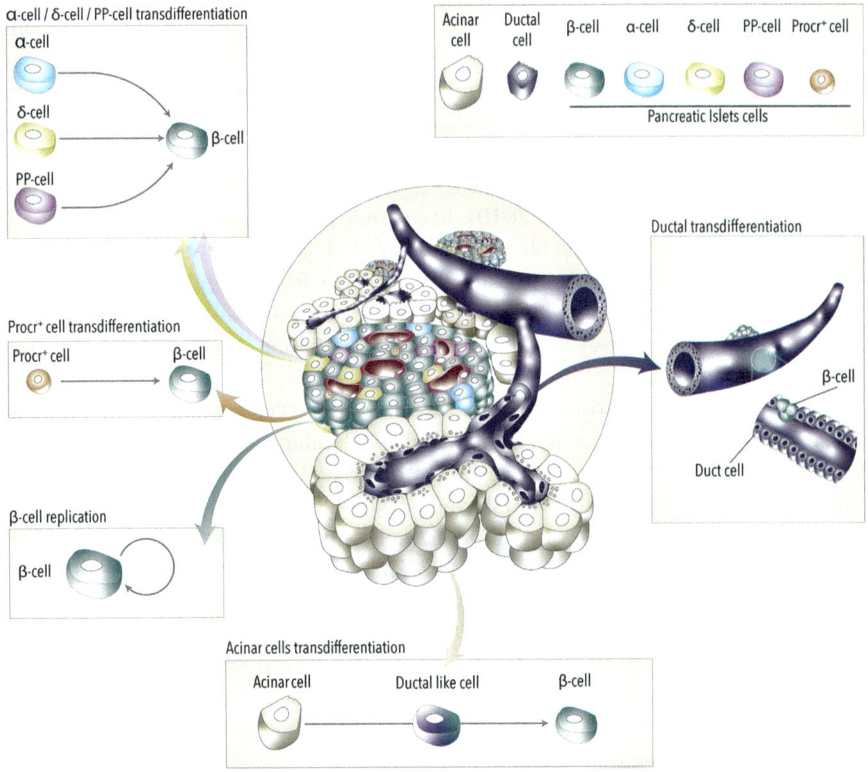

Fig. 1 Pancreas plasticity and cellular interconversions. The pancreas is composed of three different cell types: acinar, ductal, and endocrine cells. Endocrine cells (α, β, δ, and PP) are located in the islets of Langerhans, including recently identified PROCR⁺ progenitor cells. Different cellular interconversions have been described to date by using animal models. In this diagram, we highlight the interconversions or transdifferentiations of terminal differentiated pancreatic lineages to β cells that have been described to date

2 β-Cell Replication

The proliferative capacity of the pancreas, as opposed to other epithelial organs, is minimal. Although islet replication is rare under normal conditions, some evidence suggests that β cells can proliferate in situations requiring increased insulin (Demcollari et al. 2017; Spears et al. 2021). Islets and β cells are substantially different between mice and humans in terms of islet distribution, composition, endocrine cell organization, vasculature, and neural innervation (Brissova et al. 2015; Kim et al. 2009; Rorsman and Ashcroft 2018). Furthermore, human β cells are more reticent to enter the cell cycle than rodent β cells (Fiaschi-Taesch et al. 2013), although some of this apparent difference stems from the age of islets typically examined in humans and rodents (Kushner 2013). In any case, the ability of β cells to replicate differs in rodents and humans, with most evidence for β-cell replication based on murine models.

One of the most substantial pieces of evidence of β-cell replication in mice has been shown during pregnancy, when maternal β cells replicate to compensate for the high insulin demand (Baeyens et al. 2016; Sorenson and Brelje 1997). In mice, maternal β-cell proliferation starts by day 9 and reaches a peak by day 12–15. This proliferation is coupled with increased β-cell survival and size and reduced apoptosis during gestation. β-cell mass expands three- to fivefold during pregnancy, but post-pregnancy, this transient expansion regresses to pre-pregnancy levels through apoptosis (Rieck and Kaestner 2010). In rodents, pregnancy-induced β-cell expansion is dependent on secreted placental lactogens via the prolactin receptor (Parsons et al. 1992). Accordingly, RNA sequencing of islets from pregnant and postpartum mice demonstrated dynamic changes in gene expression patterns, showing robust downregulation of cell-cycle-related genes in maternal islets 1 day after birth (Takahashi et al. 2020).

While plenty of experimental evidence exists for β-cell replication during pregnancy in mice, information about maternal β-cell adaptation during human gestation is relatively limited. In accordance with murine β cells, maternal β-cell mass in humans increases by 1.4–2.4-fold during gestation; however, the authors could not detect any increase in the proliferation marker Ki67 when comparing pregnant and non-pregnant women (Butler et al. 2010). This lack of change could result from a proliferation peak too narrow to be captured on the limited number of human samples examined. In addition, since gestation time in humans lasts 13 times longer than in mice, one could expect a lower human β-cell proliferation rate at any given time point.

Mouse lineage-tracing techniques have been fundamental to understanding and uncovering the replication potential of β cells after different challenges (Dor et al. 2004). In these models, β cells are irreversibly labeled using mouse reporter lines and followed over time in homeostatic and challenging conditions. In addition to pregnancy, partial pancreatectomy in mice leads to a rapid β-cell mass expansion (Peshavaria et al. 2006). Although several mechanisms have been suggested to contribute to β-cell proliferation, lineage-tracing experiments using *RIP-Cre*^{ER}*;Z/AP* have demonstrated that pre-existing β cells are the main source of new β cells after pancreatectomy in adult mice (Dor et al. 2004). Furthermore, the loss of β cells following transient diphtheria toxin expression in adult β cells reverses through the replication of surviving β cells (Nir et al. 2007). This result suggests that mouse β cells in young adults retain their replication potential in vivo.

Follow-up studies using similar lineage-tracing strategies and more sophisticated lineage-tracing methods further confirmed that β-cell replication is the main mechanism to achieve islet expansion during pregnancy and after partial pancreatectomy (Zhao et al. 2021). However, by 8 months of age, pregnancy no longer provokes increased β-cell replication (Rankin and Kushner 2009). By 12 months of age, proliferation in response to partial pancreatectomy was severely reduced in mice compared to those at 8 weeks; furthermore, partial pancreatectomy at 14 months of age did not induce β-cell replication, indicating that murine β cells lose their replicative potential with age (Rankin and Kushner 2009). In contrast to mice up to 12 months of age, no β-cell regeneration has been observed in adult humans who had undergone partial pancreatectomy (Menge et al. 2008). However, the mean age of the subjects in this study was 51.9, with a range from 39 to 72. This older age in human subjects may be a partial reason for the lack of increased replication. In addition, the mean time between the initial partial pancreatectomy and the later tissue collection was 1.8 years, ranging between 8 days and 3.5 years (Menge et al. 2008). The follow-up time may have exceeded the period in which any β-cell regeneration occurred.

Changes in metabolic demands also play a role in β-cell replication. Proliferation of β cells has been observed under metabolic stress or in mice fed a high-fat diet (Cox et al. 2016; Kahn et al. 2006). These findings are supported by evidence in humans, wherein obesity has been reported as a trigger to increase β-cell replication (Linnemann et al. 2014). In mice, a fasting-mimicking diet is sufficient to induce an embryonic gene expression program in adult β cells, resulting in NGN3-mediated β-cell neogenesis. Interestingly, reversion of β-cell failure in T1D and T2D mouse models was achieved after repetitive cycles of fasting (Cheng et al. 2017).

The signaling pathways that regulate β-cell proliferation have been a matter of interest for decades due to their potential use for expanding β cells in vivo or ex vivo. Key cell-cycle regulators and circulating factors control β-cell proliferation. Multiple mitogenic pathways, including IRS-PI3K-AKT, RAS/RAF/ERK, mTOR, and the calcium-dependent activated proliferating transcription factor NFAT, mediate β-cell replication (Stewart et al. 2015). Upstream activators of these pathways include nutrients (glucose), growth factors (GLP1, GIP), and hormones (prolactin, progesterone, and leptin, among others). The activation of such pathways results in β-cell cell-cycle re-entry via the concerted action of cyclins, CDKs, cell-cycle inhibitors, and E2F factors (Stewart et al. 2015). Concordantly, mitogenic stimuli such as Exendin-4 and GLP1 trigger β-cell cell-cycle re-entry by their combined action on Cyclin A, CDK1, and NFAT (Lavine and Attie 2010; Xu et al. 1999). However, replicating β cells adopt transcriptomic and metabolic profiles that resemble those of immature neonatal β cells, which harbor an excellent replication potential (Puri et al. 2018). Conversely, induction of β-cell cell-cycle re-entry by proliferative cues promotes a functionally immature phenotype. Interestingly, dedifferentiation of β cells to immature NGN3-expressing β cells has been reported under glucotoxic conditions. This phenotype reverts when glucose levels are restored (Wang et al. 2014). Furthermore, dedifferentiation of mature insulin-producing cells to a naive status has been reported as a novel mechanism of T2D β-cell failure (Talchai et al. 2012). These data suggest that modulating the mitogenic pathways in β cells is a good approach to expand the β-cell pool in vivo or ex vivo.

The ability of younger animals to regenerate tissue has been demonstrated for years (López-Otín et al. 2013). β-cell maturation continues after birth for a few weeks, when they display a high proliferative capacity (Otonkoski et al. 1988, 1991; Bliss and Sharp 1992). β-cell replication potential is evident in fetal and neonatal stages but rapidly declines with age (Tschen et al. 2009; De Tata 2014). Moreover, studies in mouse models have shown that β cells can trigger compensatory proliferation after induction of β-cell-specific ablation and hyperglycemia (Nir et al. 2007). Interestingly, this dynamic proliferative response is mediated by signals derived from increased β-cell aerobic glucose metabolism, the same process that promotes other glucose-dependent responses such as insulin secretion (Porat et al. 2011). Surprisingly, analysis of young β-cell replication in response to hyperglycemia that occurred during a series of experiments in which the mice were prematurely weaned on different diets revealed that compensatory proliferation is a trait of mature β cells. β-cell compensation requires a discrete maturation step associated with the dietary change that occurs at weaning. These data show that the change from intake of fatty-acid-rich maternal milk to a carbohydrate-rich diet enables glucose-induced responsiveness (Stolovich-Rain et al. 2015).

One open question in the field is whether all the β cells have the same potential to proliferate or if there is a dedicated pool that specializes in replication upon specific triggers. β cells show substantial heterogeneity in cell complexity in granules, glucose responsiveness, insulin processing, and transcriptomic levels (Benninger and Kravets 2022). Initial mouse lineage-tracing and BrdU/Idu-labeling experiments suggested that all islet β cells have equal proliferative potential (Dor et al. 2004; Brennand and Melton 2009; Teta et al. 2007). More recently, single-cell transcriptomics has been extensively used to further understand the different types of β cells and to infer cues about their function and replication potential (Muraro et al. 2016; Wang et al. 2020).

Interestingly, Protein C Receptor $(PROCR)^+$ cells identified by scRNA-seq analysis of mouse islets may be able to differentiate into new α, β, and δ cells during homeostasis (Wang et al. 2020). Flattop (Fltp1) expression has also been suggested as a marker that functionally segregates β cells. Fltp1-Venus reporter expression demonstrated that Fltp1 subdivides β cells into two populations and distinguishes replication-competent from mature β cells (Bader et al. 2016). In more recent work, scRNAseq analysis of islets isolated from control and pancreatectomized mice identified four different β-cell clusters, one of which showed a highly proliferative signature. Interestingly, pseudo-temporal analysis suggested a sequential transition from non-proliferating to proliferating clusters (Tatsuoka et al. 2020). However, the findings in this work could also be consistent with a stochastic model of β-cell replication.

Nevertheless, the replicative capacity of β cells in response to several different stimuli is apparent. Furthermore, replication is still considered to account for most of the β-cell regeneration that occurs during normal pancreatic homeostasis. Although changes in β-cell mass and turnover in rodents are evident, the results are more limited in humans. Further analysis is required to determine to what extent the differences between rodent and human β-cell replication reflect actual differences between the species versus limitations in experimental design inherent in studying the human pancreas.

3 Intra-islet Plasticity

Although sparse β-cell replication explains their limited regeneration, different mouse lineage-tracing models have demonstrated that upon severe β-cell ablation, new insulin-producing β cells are generated from the conversion of pancreatic α or δ cells in adult and adolescent mice (Chera et al. 2014; Thorel et al. 2010). Though this is an exciting therapeutic avenue to explore for T1D patients, the existence of these intra-islet conversions in humans remains unclear.

3.1 α- to β-Cell Conversion

Adult glucagon-producing α cells and insulin-producing β cells are both generated from a common NGN3-expressing progenitor during embryonic development (Pan and Wright 2011; Edlund 2002; Gradwohl et al. 2000). The coordination of transcriptions factors is the key determinant for α- and β-cell fate commitment. One of the fundamental axes operating in the dichotomous β-to-α cell specification during embryogenesis is formed by the PAX4 and ARX transcription factors. While ARX activation drives an α-cell fate and represses PAX4, PAX4 promotes a β-cell fate and represses the α-cell program (Collombat et al. 2003). After this initial specification, further maturation of α and β cells into their final fate is achieved by differential activation of transcription factors. In this process, MAFB and IRX2 guide α-cell maturation and PDX1, MAFA, and NKX6.1 drive β-cell maturation (Collombat et al. 2003; Qiu et al. 2017). Adult α- and β-cell identity is set by the unique transcriptional networks of each differentiated cell type.

Various experimental approaches in mice have demonstrated that the identity of terminally differentiated cells in adult islets can be modified. Murine experiments combining conditional knockout and lineage-tracing technology have demonstrated that both ARX and DNMT1 are critical for the maintenance of α-cell identity. Supporting this conclusion, specific co-deletion of both genes in adult mice results in an extensive α-to-β cell conversion (Chakravarthy et al. 2017). In line with this result, forced overexpression of the α-cell fate repressor PAX4, specifically in murine α cells, leads to α-to-β cell conversion. Extensive lineage tracing of the different pancreatic cell types demonstrates that PAX4-induced α-to-β cell conversion is accompanied by mobilization of duct-lining cells that proliferate and restore the α-cell pool (Collombat et al. 2009).

Interestingly, the process of α-to-β cell conversion has been recapitulated in mice treated with an ARX inhibitor, gamma-aminobutyric acid (GABA) (Ben-Othman et al. 2017). Since GABA is an abundant neurotransmitter in the mature brain, this result could suggest that physiological processes that modulate GABA levels could activate the α-to-β cell transition transcriptional program. However, it is not clear whether the GABA levels achieved upon physiological stimuli are similar to those used in the described mouse model. In addition, several other publications could not replicate this result (Ackermann et al. 2018; Yi et al. 2020; Gu et al. 2023; Shin et al. 2019; Townsend and Skehel 1984).

In agreement with the notion that islet cell identity can be experimentally altered, deletion of XBP1 (endoplasmic reticulum stress response factor C-box binding protein) in adult mouse β cells led to β-to-α cell transdifferentiation and increased α-cell mass (Below et al. 2022). XBP1 deletion decreased β-cell distinctive genes such as *INS*, *PDX1*, *NKX6.1*, *BETA2*, and *forkhead box protein (FOX)O1*, while it induced the de-repression of β-cell dedifferentiation (*ALDH1A3*) and α-cell identity (glucagon, *ARX*, *IRX2*) genes. Furthermore, deletion of *XBP1* in $Lepr^{ob/ob}$ or high-fat diet-treated mice induced diabetes by disrupting insulin secretion (Below et al. 2022).

Other overexpression strategies that have uncovered α-to-β cell plasticity involved the in vivo overexpression of *Pdx1* and *Mafa* via adenoviral infusion in the pancreatic duct of mice. This process led to sufficient reprogramming of α-to-β cells to normalize blood glucose levels for up to 4 months in non-obese diabetic (NOD) and STZ/Alloxan-treated mice (Xiao et al. 2018).

In addition to overexpression strategies involving factors that either promote β-cell differentiation or repress α-cell identity, intercellular conversions within the islets are observed when extreme β-cell depletion is chemically or genetically induced in rodents. Transgenic expression of diphtheria toxin resulting in complete β-cell ablation combined with insulin administration results in ~1% of α cells transdifferentiating to β cells in the absence of proliferation (Thorel et al. 2010). Though the generated β cells are fully functional, the permissive window for α-to-β cell reprogramming is restricted to mice above 6 weeks of age. Interestingly, other reports have demonstrated that in younger animals (pre-puberty), δ cells, but not α cells, transdifferentiate to β cells upon extreme β-cell loss (Chera et al. 2014). This data suggests that distinct temporal windows allow α- or δ-cell plasticity to exist in mice. Interestingly, β-cell regeneration has been observed in children with T1D (Karges et al. 2004), although it is unclear whether the regenerative mechanism involves α- or δ-to-β-cell conversion or β-cell replication.

Furthermore, α-to-β-cell conversion may occur during homeostasis in healthy islets without β-cell depletion triggers. Under basal conditions, a population of cells in an intermediate stage in the transdifferentiation from α-to-β cells located in the periphery of the islets in a "neogenic niche" was identified. These cells, termed "Virgin" β cells by the authors, were characterized as being transcriptionally and functionally immature β cells lacking Urocortin-3 (UCN3) expression (van der Meulen et al. 2017).

3.2 δ- to β-Cell Conversion

Like α and β cells, δ cells are derived from a pan-endocrine progenitor expressing NGN3 during embryonic development. Thus, the possibility of a δ-to-β-cell conversion has also been explored. A decade ago, the Herrera laboratory highlighted the potential of δ-to-β-cell conversion. By completely ablating β cells in 2-week-old RIP-DTR mice with concomitant lineage tracing of δ cells (*Somatostatin(SST)-Cre;R26-YFP*), they showed that prepubescent mice rapidly recovered from diabetes after near-total β-cell loss. Under these settings, no α-to-β-cell conversion was observed; instead, δ cells dedifferentiated, proliferated, and re-expressed islet developmental genes in a FOXO1-dependent process (Chera et al. 2014) before giving rise to β cells. Significantly, δ-to-β cell conversion was highly efficient, as half of the dedifferentiated δ-cell progeny became insulin-positive, and bihormonal SST^{+}/Ins^{+} cells were rare.

In the same mouse lines, Druelle et al. demonstrated that ectopic expression of PAX4 in δ cells is sufficient to induce their conversion into functional β-like cells. Importantly, this conversion causes compensatory mechanisms involving the reactivation of endocrine developmental processes, which results in dramatic hyperplasia of the β-like cells. Moreover, ectopic PAX4 expression in δ cells induced ductal-cell proliferation and reawakening of endocrine developmental processes, characterized by re-expression of NGN3 in duct-lining cells. Renewed endocrine development was accompanied by its associated developmental processes, such as epithelial-to-mesenchymal transition markers, and these cells eventually adopted an endocrine cell identity. Indeed, although no ductal lineage tracing was analyzed in this study, the data suggested that proliferating ductal cells eventually adopt an endocrine cell identity by reenacting the endocrine developmental program (Druelle et al. 2017).

Finally, the Manfroid laboratory recently reported that a specific δ-cell subpopulation distinct from the previously identified zebrafish Sst2 δ cells and characterized by the expression of Sst1.1 displays several important β-cell features. The transcriptomic profile of bihormonal cells is also very close to that of the $Sst1.1^{+}$ δ cells, making them resemble β/δ hybrids. In vivo imaging of larvae revealed the emergence of insulin-expressing bihormonal cells from monohormonal sst1.1 δ cells early after β-cell ablation (Carril Pardo et al. 2022). They also provide evidence that pancreatic ducts contribute to the pool of bihormonal cells. Furthermore, bihormonal cells were abundant in the regenerated pancreas and able to normalize glycemia after a glucose challenge.

3.3 Pancreatic Polypeptide to β Conversion

One of the rarest endocrine cell types is pancreatic polypeptide (PPY)-expressing γ cells. Recent work highlighted subsets of γ cells that produce hormones in addition to PPY. These bihormonal cells share identity markers typical of the other islet cell types. In mice, *Ppy* gene inactivation or conditional γ-cell ablation did not alter glycemia or body weight. Interestingly, upon β cell injury, γ cells exhibited gene expression changes, and some of them engaged in insulin production, as previously observed for α and δ cells (Perez-Frances et al. 2021). These results highlight γ-cell plasticity and suggest further investigation as an alternative source of β cells.

3.4 PROCR$^+$ Cells as Progenitors Residing in the Islets

scRNA technologies allow for the characterization of endocrine heterogeneity and the identification in mice of a new cell population between the islets characterized by the expression and the absence of endocrine differentiation markers (Wang et al. 2020), representing 1% of islet cells. Lineage-tracing studies demonstrated that PROCR$^+$ cells contribute to all endocrine cell progeny and form islet-like organoids in culture that can differentiate into endocrine lineages; within these differentiated organoids, β cells dominate, while α, δ, and PP cells occur at lower frequencies. Moreover, upon transplantation in diabetic mice, these PROCR-derived organoids reverse disease, suggesting that PROCR$^+$ cells represent endocrine progenitors in the adult pancreas (Wang et al. 2020, 2022). However, the contribution of the PROCR$^+$ population to endocrine regeneration has not yet been studied.

4 Exocrine-to-β-Cell Plasticity

The exocrine pancreas comprises over 95% of the gland. Acinar cells are the most abundant exocrine cells, while ductal cells represent approximately 20% of the exocrine tissue. Thus, these cell types represent both by lineage proximity to β cells and by abundance an interesting starting material to investigate for the capacity to give rise to β cells.

4.1 Ductal to β-Cell Conversion

All pancreatic epithelial cells derive from a multipotent progenitor cell. This type of progenitor gives rise to the acinar progenitors and bipotent progenitors that appear around the time of the secondary transition. Ductal and endocrine cells arise from the bipotent progenitor, which expresses SOX9, HNF1β, and KRT19. These proteins are classical ductal markers in adult tissue (Pan et al. 2013; Zhou et al. 2007; Solar et al. 2009). Upon Notch pathway inhibition, the bipotent progenitor activates the expression of the endocrine progenitor marker NGN3, thus activating an endocrine fate (Gradwohl et al. 2000). By default, those bipotent progenitors that maintain active Notch will downregulate the pancreatic progenitor marker PDX1 and differentiate into mature pancreatic duct cells with sustained expression of SOX9, HNF1β, and HNF6 (Kopp et al. 2011; Seymour 2014).

Duct and endocrine cell proximity has been observed during normal growth and is even more prominent during pregnancy and regeneration (Xu et al. 1999, 2008; Gu and Sarvetnick 1993; Rosenberg 1995; Rooman and Bouwens 2004; Li et al. 2010; Bonner-Weir et al. 1993; Yoneda et al. 2013; El-Gohary et al. 2012). The close apposition of ducts and islets—with endocrine cells even embedded within or appearing to bud from ducts in the mammalian pancreas (Xu et al. 1999, 2008; Gu and Sarvetnick 1993; Rosenberg 1995; Rooman and Bouwens 2004; Li et al. 2010; Bonner-Weir et al. 1993; Yoneda et al. 2013; El-Gohary et al. 2012), together with a common progenitor for ductal and endocrine cells, has inspired researchers to explore the existence of putative or facultative progenitors residing in the ducts. Multiple lineage-tracing studies, however, have conflicting results (Afelik and Rovira 2017).

Initially, several studies suggested that adult pancreatic duct cells contribute to pancreatic β-cell regeneration after pancreatic duct ligation (PDL). PDL is an injury model that causes the destruction of the acinar tissue in the tail of the pancreas by blocking the main pancreatic duct. Under this condition, NGN3 is reactivated in ductal cells, reminiscent of the developing pancreas, and β-cell mass is increased by twofold (Xu et al. 2008). During the same period, Inada et al. used the human *carbonic anhydrase II* (*CAII*) promoter to derive *CAII-Cre*^{ER} mice for lineage tracing. Analysis of this transgenic mouse model suggested that CAII-expressing cells serve as progenitors that contribute to both exocrine and endocrine lineages in homeostatic conditions and following injury by PDL (Inada et al. 2008). Criscimanna et al. also reported that when acinar and endocrine tissues are selectively ablated by exposure of *Pdx1-Cre;R26DTR* mice to diphtheria toxin, the surviving ductal cells contributed to both endocrine and acinar regeneration (Criscimanna et al. 2011). Finally, a recent study identified NGN3-expressing cells in the duct as a source of adult β-cell neogenesis in homeostasis and diabetes, suggesting that this mechanism, in addition to β-cell proliferation, maintains the adult islet β-cell population. Whether these NGN3 cells are retained in the ductal epithelium from development or are a new subpopulation of ductal cells in the adult requires further investigation (Gribben et al. 2021).

Contradicting the above studies, many other mouse lineage-tracing studies using various adult pancreatic duct-specific promoters, including HNF1β, Mucin1, SOX9, and HEX1, have not observed any contribution from ductal cells to β-cell regeneration (Solar et al. 2009; Kopp et al. 2011; Kopinke et al. 2011; Kopinke and Murtaugh 2010; Furuyama et al. 2011), even under injury-induced regenerating conditions

(Solar et al. 2009; Kopp et al. 2011; Kopinke et al. 2011; Kopinke and Murtaugh 2010). In these studies, duct cells contributed to the endocrine pancreas lineage only during embryonic and, in some cases, prepubescent stages, but not in the adult. Although the induction of NGN3 by PDL was noted in some of these subsequent studies, as previously reported, these NGN3-expressing duct cells did not contribute to new β-cell formation (Kushner et al. 2010).

While this conflicting evidence remains unresolved, we can conclude that no dedicated progenitors residing in the ductal epithelium have been identified. Still, the evidence suggests that pancreatic duct cells can be reprogrammed into β cells in vivo by genetic manipulation of transcription factors or signaling pathways. Al-Hasani et al. recently described how $NGN3^+$ cells lining the ducts contribute to β-cell expansion induced by overexpressing PAX4 in α cells (Al-Hasani et al. 2013). Moreover, ductal cells are capable of converting into α cells, δ cells, and β cells following the inactivation of the SCF-type E3 ubiquitin ligase substrate component FBW7. Deletion of FBW7 stabilizes NGN3 and seems to reactivate an endocrine developmental differentiation program in adult pancreatic ductal cells (Sancho et al. 2014). El-Gohary et al. also observed that β cells arise from specialized ductal structures called intra-islet ducts following partial pancreatectomy in adult TGF-β receptor mutant mice (El-Gohary et al. 2016). Recently, lineage-tracing studies using *Sox9-Cre*^{ER} transgenic mice have shown that adult pancreatic ductal cells can differentiate into β cells under specific conditions, which include moderate hyperglycemia and long-term administration of low-dose gastrin and epidermal growth factor (Zhang et al. 2016).

Additionally, inducible misexpression of NGN3 in ductal cells in vivo induced islet hypertrophy attributed to a proportional increase in all endocrine cell populations. In this model, lineage-tracing experiments indicated a continuous neogeneration of endocrine cells exhibiting a ductal ontogeny (Vieira et al. 2018). Finally, subpopulations of duct cells expressing high levels of SOX9, HNF1β, and CD24 (mouse) or CD51 (human) expanded in vitro and differentiated into endocrine progenitors through the application of embryonic stem cell differentiation protocols (Rezanejad et al. 2018).

The progenitor potential of centroacinar cells/terminal ductal cells (CACs/TDCs) has also been of great interest in the past few years. The capacity of CACs/TDCs to proliferate in response to pancreatic injury has been well described previously (Hayashi et al. 1999, 2003; Gasslander et al. 1992). Also, CACs/TDCs maintain active Notch signaling, reminiscent of embryonic progenitor cells. So far, in mammals, only HES1 and ALDH1A1 and ALDH1B1 have emerged as specific markers of CACs/TDCs (Kopinke et al. 2011; Rovira et al. 2010; Ioannou et al. 2013; Mameishvili et al. 2019); however, in lineage-tracing studies using *Hes1-Cre*^{ER}, no contribution of CACs/TDCs to β-cell regeneration was observed (Solar et al. 2009; Kopp et al. 2011; Kopinke et al. 2011; Kopinke and Murtaugh 2010). Adding to the controversy, CACs have recently been shown, by two independent laboratories, to be a source of β-cell neogenesis in the adult zebrafish pancreas, mirroring their progenitor function during development (Delaspre et al. 2015; Ghaye et al. 2015). Interestingly, adult zebrafish CACs retain expression of Nkx6.1 and Nkx2.2a, which are not found in mammalian CACs, but are expressed in mammalian islet cells and are master regulators of the endocrine differentiation program (Sussel et al. 1998; Schaffer et al. 2011, 2013; Sander et al. 2000; Binot et al. 2010; Papizan et al. 2011). Thus, the enhanced regenerative capacity of CACs in zebrafish might be explained by the retained expression of Nkx6.1 and Nkx2.2.

Ductal cells have been poorly characterized by scRNAseq technologies, only representing a few thousand cells among many collected. In humans, Qadir et al. identified a subpopulation expressing *PDX1* and *BMPR1a* (ALK3) in the central pancreatic ducts that expanded in response to BMP7 and differentiated upon BMP7 withdrawal. These cells were further characterized by scRNAseq, identifying $ALK3^+/P2RY1^+$ (surrogate of PDX1) cells as a subpopulation of $PDX1^+/ALK3^+/CAII^-$ cells able to differentiate into endocrine and exocrine lineages upon transplantation in immunodeficient mice (Qadir et al. 2020). However, a proof of concept of the contribution of such subpopulations to the endocrine compartment during homeostasis or recovery in vivo will be necessary. In mice, Handley et al. have recently characterized the coexistence of four ductal populations by scRNAseq, although the role of these cells in β-cell regeneration has not been studied (Hendley et al. 2021).

Finally, pancreatic murine and mature human ductal cells are able to form organoids in 3D cultures (Huch et al. 2013; Hindley et al. 2016; Boj et al. 2015; Huch and Koo 2015; Loomans et al. 2018; Broutier et al. 2016). This ability is unique to adult progenitor cells, suggesting that ductal cells display specific progenitor properties in culture. Several reports have also demonstrated that these organoids can differentiate into insulin-producing cells in vitro and in vivo under the kidney capsule, although with limited efficiency (Loomans et al. 2018; Azzarelli et al. 2018, 2017; Rezanejad et al. 2019).

4.2 *Acinar-to-β-Cell Conversion*

Acinar cells are the most abundant pancreatic cell type and therefore constitute an attractive source of reprogrammable tissue that could generate β cells, although at the developmental level, they are the furthest from the endocrine lineages (Afelik and Rovira 2017).

Pancreatic acinar cells are highly plastic. Although during normal homeostasis, acinar and ductal cells do not interconvert, during injury recovery, such as pancreatitis, some acinar cells transdifferentiate into a ductal-like or progenitor-like state, with a concomitant loss of digestive enzyme expression and activation of classical ductal markers, such as KRT19, SOX9, and HES1. In vitro, mouse acinar cells in suspension cultures also transdifferentiate or dedifferentiate into ductal-like cells (Rooman et al. 2000) but, when cultured with the growth factor EGF and leukemia inhibitory factor, newly formed β cells appear and restore normoglycemia in alloxan-induced diabetic mice (Baeyens et al. 2005). Further work showed that inhibiting Notch signaling led to a 30% conversion of acinar cells to β cells (Baeyens et al. 2009). Some of these findings were replicated ex vivo using human pancreatic tissue (Lemper et al. 2015; Houbracken et al. 2011).

In 2008, Zhou et al. reported that upon in vivo ectopic expression of master regulators of endocrine differentiation (NGN3, PDX1, and MAFA), acinar cells reprogrammed directly into insulin-producing β cells, without cell replication or reversion into a ductal-like stage (Zhou et al. 2008). Further studies demonstrated that NGN3 promotes the establishment of a generic endocrine state in acinar cells and promotes δ-cell specification in the absence of other factors. δ-cell specification is, in turn, suppressed by MAFA and PDX1 during α- and β-cell induction. Thus, these experiments identified a set of defined factors whose combinatorial actions reprogram acinar cells to distinct islet endocrine subtypes in vivo (Li et al. 2014). Acinar-derived β-like cells ameliorated blood glucose levels in diabetic mice, although this conversion took place without activating an epithelial-to-mesenchymal transition program. Therefore, newly formed β-like cells did not cluster to resemble islets, and their mature phenotype and function remain unclear. Recently, though, Miyazaki et al. demonstrated that expression of PDX1 in acinar cells was sufficient to reprogram acinar cells into endocrine precursor cells that migrated into islets and differentiated into β, δ, or PP cells (Miyazaki et al. 2016).

Finally, lineage-tracing strategies to investigate acinar-to-endocrine cell conversion using acinar-specific promoters have been developed. However, they have so far yielded conflicting results. Lineage-tracing of acinar cells using *Elastase-Cre*^{ER} showed no evidence of acinar-to-β-cell conversion during the regenerative response to injury by partial pancreatectomy, pancreatic duct ligation (PDL), or caerulein-induced pancreatitis (Desai et al. 2007). In contrast, Pan et al. demonstrated by lineage-tracing acinar cells using *Ptf1a-Cre*^{ER} transgenic mice that acinar cells can transdifferentiate into ductal cells and β cells upon pancreatic duct ligation without ectopic expression of transcription factors or cytokine stimulation, albeit at low efficiency, as only ~2% of the endocrine cells were derived from acinar cells (Pan et al. 2013).

5 Concluding Remarks

To date, no dedicated β-cell progenitors have been identified in the adult pancreas. However, evidence accumulated in the last few decades has demonstrated apparent cell fate plasticity within the pancreas. Terminally differentiated lineages, when exposed to specific injury models, stimulation with particular growth factors/cytokines, or overexpression/misexpression of key transcription factors, can transdifferentiate into insulin-producing β cells, although at low efficiency. Further investigation of efficient transdifferentiating stimuli will be vital to induce β-cell regeneration from pancreatic cells. In addition, state-of-the-art single-cell techniques could highlight the existence of cell states or cell types more capable of giving rise to β cells.

Acknowledgments This work is supported by King's College London and MRC (MR/S000011/1 and MR/T015470/1) to RS and by Ministerio de Ciencia, Innovación y Universidades (PID2022-136973OB-I00 funded by MCIN/AEI/10.13039/501100011033 and RYC-2017-21950) to MR and by PRE2020-092678 to AH. We thank CERCA Programme/Generalitat de Catalunya for institutional suport. We would like to thank and acknowledge Mateo Casas (mcasasengel@gmail.com) for the illustration in the manuscript.

Compliance with Ethical Standards Rocio Sancho declares that she has no conflict of interest. Meritxell Rovira declares that she has no conflict of interest. Adrián Holguín declares that he has no conflict of interest. This chapter is a review of previously published accounts; as such, no animal or human studies were performed.

References

Ackermann AM, Moss NG, Kaestner KH (2018) GABA and artesunate do not induce pancreatic alpha-to-beta cell transdifferentiation in vivo. Cell Metab 28(5):787–792 e783

Afelik S, Rovira M (2017) Pancreatic β-cell regeneration: facultative or dedicated progenitors? Mol Cell Endocrinol 445:85–94

Ahlqvist E, Prasad RB, Groop L (2020) Subtypes of type 2 diabetes determined from clinical parameters. Diabetes 69(10):2086–2093

Al-Hasani K, Pfeifer A, Courtney M, Ben-Othman N, Gjernes E, Vieira A, Druelle N, Avolio F, Ravassard P, Leuckx G, Lacas-Gervais S, Ambrosetti D, Benizri E, Hecksher-Sorensen J, Gounon P, Ferrer J, Gradwohl G, Heimberg H, Mansouri A, Collombat P (2013) Adult duct-lining cells can reprogram into β-like cells able to counter repeated cycles of toxin-induced diabetes. Dev Cell 26(1):86–100

Azzarelli R, Hurley C, Sznurkowska MK, Rulands S, Hardwick L, Gamper I, Ali F, McCracken L, Hindley C, McDuff F, Nestorowa S, Kemp R, Jones K, Göttgens B, Huch M, Evan G, Simons BD, Winton D, Philpott A (2017) Multi-site neurogenin3 phosphorylation controls pancreatic endocrine differentiation. Dev Cell 41(3):274–286.e275

Azzarelli R, Rulands S, Nestorowa S, Davies J, Campinoti S, Gillotin S, Bonfanti P, Göttgens B, Huch M, Simons B, Philpott A (2018) Neurogenin3 phosphorylation controls reprogramming efficiency of pancreatic ductal organoids into endocrine cells. Sci Rep 8(1):15374

Bader E, Migliorini A, Gegg M, Moruzzi N, Gerdes J, Roscioni SS, Bakhti M, Brandl E, Irmler M, Beckers J, Aichler M, Feuchtinger A, Leitzinger C, Zischka H, Wang-Sattler R, Jastroch M, Tschöp M, Machicao F, Staiger H, Häring HU, Chmelova H, Chouinard JA, Oskolkov N, Korsgren O, Speier S, Lickert H (2016) Identification of proliferative and mature β-cells in the islets of Langerhans. Nature 535(7612):430–434

Baeyens L, De Breuck S, Lardon J, Mfopou JK, Rooman I, Bouwens L (2005) In vitro generation of insulin-producing beta cells from adult exocrine pancreatic cells. Diabetologia 48(1):49–57

Baeyens L, Bonné S, Bos T, Rooman I, Peleman C, Lahoutte T, German M, Heimberg H, Bouwens L (2009) Notch signaling as gatekeeper of rat acinar-to-beta-cell conversion in vitro. Gastroenterology 136(5):1750–1760.e1713

Baeyens L, Hindi S, Sorenson RL, German MS (2016) β-cell adaptation in pregnancy. Diabetes Obes Metab 18(Suppl 1):63–70

Bastidas-Ponce A, Scheibner K, Lickert H, Bakhti M (2017) Cellular and molecular mechanisms coordinating pancreas development. Development 144(16):2873–2888

Below CR, Kelly J, Brown A, Humphries JD, Hutton C, Xu J, Lee BY, Cintas C, Zhang X, Hernandez-Gordillo V, Stockdale L, Goldsworthy MA, Geraghty J, Foster L, O'Reilly DA, Schedding B, Askari J, Burns J, Hodson N, Smith DL, Lally C, Ashton G, Knight D, Mironov A, Banyard A, Eble JA, Morton JP, Humphries MJ, Griffith LG, Jørgensen C (2022) A microenvironment-inspired synthetic three-dimensional model for pancreatic ductal adenocarcinoma organoids. Nat Mater 21(1):110–119

Benninger RKP, Kravets V (2022) The physiological role of β-cell heterogeneity in pancreatic islet function. Nat Rev Endocrinol 18(1):9–22

Ben-Othman N, Vieira A, Courtney M, Record F, Gjernes E, Avolio F, Hadzic B, Druelle N, Napolitano T, Navarro-Sanz S, Silvano S, Al-Hasani K, Pfeifer A, Lacas-Gervais S, Leuckx G, Marroquí L, Thévenet J, Madsen OD, Eizirik DL, Heimberg H, Kerr-Conte J, Pattou F, Mansouri A, Collombat P (2017) Long-term GABA administration induces alpha cell-mediated beta-like cell neogenesis. Cell 168(1–2):73–85.e11

Binot AC, Manfroid I, Flasse L, Winandy M, Motte P, Martial JA, Peers B, Voz ML (2010) Nkx6.1 and nkx6.2 regulate alpha- and beta-cell formation in zebrafish by acting on pancreatic endocrine progenitor cells. Dev Biol 340(2):397–407

Bliss CR, Sharp GW (1992) Glucose-induced insulin release in islets of young rats: time-dependent potentiation and effects of 2-bromostearate. Am J Phys 263(5 Pt 1):E890–E896

Boj SF, Hwang CI, Baker LA, Chio II, Engle DD, Corbo V, Jager M, Ponz-Sarvise M, Tiriac H, Spector MS, Gracanin A, Oni T, Yu KH, van Boxtel R, Huch M, Rivera KD, Wilson JP, Feigin ME, Öhlund D, Handly-Santana A, Ardito-Abraham CM, Ludwig M, Elyada E, Alagesan B, Biffi G, Yordanov GN, Delcuze B, Creighton B, Wright K, Park Y, Morsink FH, Molenaar IQ, Borel Rinkes IH, Cuppen E, Hao Y, Jin Y, Nijman IJ, Iacobuzio-Donahue C, Leach SD, Pappin DJ, Hammell M, Klimstra DS, Basturk O, Hruban RH, Offerhaus GJ, Vries RG, Clevers H, Tuveson DA (2015) Organoid models of human and mouse ductal pancreatic cancer. Cell 160(1–2):324–338

Bonner-Weir S, Baxter LA, Schuppin GT, Smith FE (1993) A second pathway for regeneration of adult exocrine and endocrine pancreas. A possible recapitulation of embryonic development. Diabetes 42(12):1715–1720

Bonner-Weir S, Sullivan BA, Weir GC (2015) Human islet morphology revisited: human and rodent islets are not so different after all. J Histochem Cytochem 63(8):604–612

Brennand K, Melton D (2009) Slow and steady is the key to beta-cell replication. J Cell Mol Med 13(3):472–487

Brissova M, Shostak A, Fligner CL, Revetta FL, Washington MK, Powers AC, Hull RL (2015) Human islets have fewer blood vessels than mouse islets and the density of islet vascular structures is increased in type 2 diabetes. J Histochem Cytochem 63(8):637–645

Broutier L, Andersson-Rolf A, Hindley CJ, Boj SF, Clevers H, Koo BK, Huch M (2016) Culture and establishment of self-renewing human and mouse adult liver and pancreas 3D organoids and their genetic manipulation. Nat Protoc 11(9):1724–1743

Butler AE, Cao-Minh L, Galasso R, Rizza RA, Corradin A, Cobelli C, Butler PC (2010) Adaptive changes in pancreatic beta cell fractional area and beta cell turnover in human pregnancy. Diabetologia 53(10):2167–2176

Carril Pardo CA, Massoz L, Dupont MA, Bergemann D, Bourdouxhe J, Lavergne A, Tarifeño-Saldivia E, Helker CS, Stainier DY, Peers B, Voz MM, Manfroid I (2022) A δ-cell subpopulation with a pro-β-cell identity contributes to efficient age-independent recovery in a zebrafish model of diabetes. Elife 11:e67576

Chakravarthy H, Gu X, Enge M, Dai X, Wang Y, Damond N, Downie C, Liu K, Wang J, Xing Y, Chera S, Thorel F, Quake S, Oberholzer J, MacDonald PE, Herrera PL, Kim SK (2017) Converting adult pancreatic islet α cells into β cells by targeting both Dnmt1 and Arx. Cell Metab 25(3):622–634

Cheng CW, Villani V, Buono R, Wei M, Kumar S, Yilmaz OH, Cohen P, Sneddon JB, Perin L, Longo VD (2017) Fasting-mimicking diet promotes ngn3-driven β-cell regeneration to reverse diabetes. Cell 168(5):775–788.e712

Chera S, Baronnier D, Ghila L, Cigliola V, Jensen JN, Gu G, Furuyama K, Thorel F, Gribble FM, Reimann F, Herrera PL (2014) Diabetes recovery by age-dependent conversion of pancreatic δ-cells into insulin producers. Nature 514(7523):503–507

Collombat P, Mansouri A, Hecksher-Sorensen J, Serup P, Krull J, Gradwohl G, Gruss P (2003) Opposing actions of Arx and Pax4 in endocrine pancreas development. Genes Dev 17(20):2591–2603
Collombat P, Xu X, Ravassard P, Sosa-Pineda B, Dussaud S, Billestrup N, Madsen OD, Serup P, Heimberg H, Mansouri A (2009) The ectopic expression of Pax4 in the mouse pancreas converts progenitor cells into alpha and subsequently beta cells. Cell 138(3):449–462
Cox AR, Lam CJ, Rankin MM, King KA, Chen P, Martinez R, Li C, Kushner JA (2016) Extreme obesity induces massive beta cell expansion in mice through self-renewal and does not alter the beta cell lineage. Diabetologia 59(6):1231–1241
Criscimanna A, Speicher JA, Houshmand G, Shiota C, Prasadan K, Ji B, Logsdon CD, Gittes GK, Esni F (2011) Duct cells contribute to regeneration of endocrine and acinar cells following pancreatic damage in adult mice. Gastroenterology 141(4):1451–1462, 1462.e1451-1456
D'Amour KA, Bang AG, Eliazer S, Kelly OG, Agulnick AD, Smart NG, Moorman MA, Kroon E, Carpenter MK, Baetge EE (2006) Production of pancreatic hormone-expressing endocrine cells from human embryonic stem cells. Nat Biotechnol 24(11):1392–1401
De Tata V (2014) Age-related impairment of pancreatic Beta-cell function: pathophysiological and cellular mechanisms. Front Endocrinol 5:138
Delaspre F, Beer RL, Rovira M, Huang W, Wang G, Gee S, MeC V, Wheelan SJ, Parsons MJ (2015) Centroacinar cells are progenitors that contribute to endocrine pancreas regeneration. Diabetes 64(10):3499–3509
Demcollari TI, Cujba AM, Sancho R (2017) Phenotypic plasticity in the pancreas: new triggers, new players. Curr Opin Cell Biol 49:38–46
Desai BM, Oliver-Krasinski J, De Leon DD, Farzad C, Hong N, Leach SD, Stoffers DA (2007) Preexisting pancreatic acinar cells contribute to acinar cell, but not islet beta cell, regeneration. J Clin Invest 117(4):971–977
Dor Y, Brown J, Martinez OI, Melton DA (2004) Adult pancreatic beta-cells are formed by self-duplication rather than stem-cell differentiation. Nature 429(6987):41–46
Druelle N, Vieira A, Shabro A, Courtney M, Mondin M, Rekima S, Napolitano T, Silvano S, Navarro-Sanz S, Hadzic B, Avolio F, Rassoulzadegan M, Schmid HA, Mansouri A, Collombat P (2017) Ectopic expression of Pax4 in pancreatic δ cells results in β-like cell neogenesis. J Cell Biol 216(12):4299–4311
Edlund H (2002) Pancreatic organogenesis—developmental mechanisms and implications for therapy. Nat Rev Genet 3(7):524–532
El-Gohary Y, Tulachan S, Branca M, Sims-Lucas S, Guo P, Prasadan K, Shiota C, Gittes GK (2012) Whole-mount imaging demonstrates hypervascularity of the pancreatic ducts and other pancreatic structures. Anat Rec 295(3):465–473
El-Gohary Y, Wiersch J, Tulachan S, Xiao X, Guo P, Rymer C, Fischbach S, Prasadan K, Shiota C, Gaffar I, Song Z, Galambos C, Esni F, Gittes GK (2016) Intraislet pancreatic ducts can give rise to insulin-positive cells. Endocrinology 157(1):166–175
Fiaschi-Taesch NM, Kleinberger JW, Salim FG, Troxell R, Wills R, Tanwir M, Casinelli G, Cox AE, Takane KK, Srinivas H, Scott DK, Stewart AF (2013) Cytoplasmic-nuclear trafficking of G1/S cell cycle molecules and adult human β-cell replication: a revised model of human β-cell G1/S control. Diabetes 62(7):2460–2470
Fuchs E (2009) Finding one's niche in the skin. Cell Stem Cell 4(6):499–502
Furuyama K, Kawaguchi Y, Akiyama H, Horiguchi M, Kodama S, Kuhara T, Hosokawa S, Elbahrawy A, Soeda T, Koizumi M, Masui T, Kawaguchi M, Takaori K, Doi R, Nishi E, Kakinoki R, Deng JM, Behringer RR, Nakamura T, Uemoto S (2011) Continuous cell supply from a Sox9-expressing progenitor zone in adult liver, exocrine pancreas and intestine. Nat Genet 43(1):34–41
Gasslander T, Ihse I, Smeds S (1992) The importance of the centroacinar region in cerulein-induced mouse pancreatic growth. Scand J Gastroenterol 27(7):564–570
Ghaye AP, Bergemann D, Tarifeño-Saldivia E, Flasse LC, Von Berg V, Peers B, Voz ML, Manfroid I (2015) Progenitor potential of nkx6.1-expressing cells throughout zebrafish life and during beta cell regeneration. BMC Biol 13(1):70

Gonzales KAU, Fuchs E (2017) Skin and its regenerative powers: an alliance between stem cells and their niche. Dev Cell 43(4):387–401

Gradwohl G, Dierich A, LeMeur M, Guillemot F (2000) Neurogenin3 is required for the development of the four endocrine cell lineages of the pancreas. Proc Natl Acad Sci USA 97(4):1607–1611

Gribben C, Lambert C, Messal HA, Hubber EL, Rackham C, Evans I, Heimberg H, Jones P, Sancho R, Behrens A (2021) Ductal Ngn3-expressing progenitors contribute to adult β cell neogenesis in the pancreas. Cell Stem Cell 28(11):2000–2008.e2004

Gu D, Sarvetnick N (1993) Epithelial cell proliferation and islet neogenesis in IFN-g transgenic mice. Development 118(1):33–46

Gu L, Cui X, Lin X, Yang J, Wei R, Hong T, Yang K (2023) Gamma-aminobutyric acid modulates alpha-cell hyperplasia but not beta-cell regeneration induced by glucagon receptor antagonism in type 1 diabetic mice. Acta Diabetol 60(1):19–28

Guiu J, Hannezo E, Yui S, Demharter S, Ulyanchenko S, Maimets M, Jørgensen A, Perlman S, Lundvall L, Mamsen LS, Larsen A, Olesen RH, Andersen CY, Thuesen LL, Hare KJ, Pers TH, Khodosevich K, Simons BD, Jensen KB (2019) Tracing the origin of adult intestinal stem cells. Nature 570(7759):107–111

Hayashi K, Takahashi T, Kakita A, Yamashina S (1999) Regional differences in the cellular proliferation activity of the regenerating rat pancreas after partial pancreatectomy. Arch Histol Cytol 62(4):337–346

Hayashi KY, Tamaki H, Handa K, Takahashi T, Kakita A, Yamashina S (2003) Differentiation and proliferation of endocrine cells in the regenerating rat pancreas after 90% pancreatectomy. Arch Histol Cytol 66(2):163–174

Hendley AM, Rao AA, Leonhardt L, Ashe S, Smith JA, Giacometti S, Peng XL, Jiang H, Berrios D, Pawlak M, Li LY, Lee J, Collisson EA, Anderson MS, Fragiadakis GK, Yeh JJ, Chun JY, Kim GE, Weaver VM, Hebrok M (2021) Single cell transcriptome analysis defines heterogeneity of the murine pancreatic ductal tree. Elife:10, e67776

Hindley CJ, Cordero-Espinoza L, Huch M (2016) Organoids from adult liver and pancreas: stem cell biology and biomedical utility. Dev Biol 420:251

Houbracken I, de Waele E, Lardon J, Ling Z, Heimberg H, Rooman I, Bouwens L (2011) Lineage tracing evidence for transdifferentiation of acinar to duct cells and plasticity of human pancreas. Gastroenterology 141(2):731–741. 741.e731-734

Huch M, Koo BK (2015) Modeling mouse and human development using organoid cultures. Development 142(18):3113–3125

Huch M, Bonfanti P, Boj SF, Sato T, Loomans CJ, van de Wetering M, Sojoodi M, Li VS, Schuijers J, Gracanin A, Ringnalda F, Begthel H, Hamer K, Mulder J, van Es JH, de Koning E, Vries RG, Heimberg H, Clevers H (2013) Unlimited in vitro expansion of adult bi-potent pancreas progenitors through the Lgr5/R-spondin axis. EMBO J 32(20):2708–2721

Inada A, Nienaber C, Katsuta H, Fujitani Y, Levine J, Morita R, Sharma A, Bonner-Weir S (2008) Carbonic anhydrase II-positive pancreatic cells are progenitors for both endocrine and exocrine pancreas after birth. Proc Natl Acad Sci USA 105(50):19915–19919

Ioannou M, Serafimidis I, Arnes L, Sussel L, Singh S, Vasiliou V, Gavalas A (2013) ALDH1B1 is a potential stem/progenitor marker for multiple pancreas progenitor pools. Dev Biol 374(1):153–163

Jonsson J, Carlsson L, Edlund T, Edlund H (1994) Insulin-promoter-factor 1 is required for pancreas development in mice. Nature 371(6498):606–609

Kahn SE, Hull RL, Utzschneider KM (2006) Mechanisms linking obesity to insulin resistance and type 2 diabetes. Nature 444(7121):840–846

Karges B, Durinovic-Belló I, Heinze E, Boehm BO, Debatin KM, Karges W (2004) Complete long-term recovery of beta-cell function in autoimmune type 1 diabetes after insulin treatment. Diabetes Care 27(5):1207–1208

Kharroubi AT, Darwish HM (2015) Diabetes mellitus: the epidemic of the century. World J Diabetes 6(6):850–867

Kim A, Miller K, Jo J, Kilimnik G, Wojcik P, Hara M (2009) Islet architecture: a comparative study. Islets 1(2):129–136
Kopinke D, Murtaugh LC (2010) Exocrine-to-endocrine differentiation is detectable only prior to birth in the uninjured mouse pancreas. BMC Dev Biol 10:38
Kopinke D, Brailsford M, Shea JE, Leavitt R, Scaife CL, Murtaugh LC (2011) Lineage tracing reveals the dynamic contribution of Hes1+ cells to the developing and adult pancreas. Development 138(3):431–441
Kopp JL, Dubois CL, Schaffer AE, Hao E, Shih HP, Seymour PA, Ma J, Sander M (2011) Sox9+ ductal cells are multipotent progenitors throughout development but do not produce new endocrine cells in the normal or injured adult pancreas. Development 138(4):653–665
Kushner JA (2013) The role of aging upon beta cell turnover. J Clin Invest 123(3):990–995
Kushner JA, Weir GC, Bonner-Weir S (2010) Ductal origin hypothesis of pancreatic regeneration under attack. Cell Metab 11(1):2–3
Lavine JA, Attie AD (2010) Gastrointestinal hormones and the regulation of β-cell mass. Ann N Y Acad Sci 1212:41–58
Lemper M, Leuckx G, Heremans Y, German MS, Heimberg H, Bouwens L, Baeyens L (2015) Reprogramming of human pancreatic exocrine cells to β-like cells. Cell Death Differ 22(7):1117–1130
Li XY, Zhan XR, Lu C, Liu XM, Wang XC (2010) Mechanisms of hepatocyte growth factor-mediated signaling in differentiation of pancreatic ductal epithelial cells into insulin-producing cells. Biochem Biophys Res Commun 398(3):389–394
Li W, Nakanishi M, Zumsteg A, Shear M, Wright C, Melton DA, Zhou Q (2014) In vivo reprogramming of pancreatic acinar cells to three islet endocrine subtypes. elife 3:e01846
Linnemann AK, Baan M, Davis DB (2014) Pancreatic β-cell proliferation in obesity. Adv Nutr 5(3):278–288
Loomans CJM, Williams Giuliani N, Balak J, Ringnalda F, van Gurp L, Huch M, Boj SF, Sato T, Kester L, de Sousa Lopes SMC, Roost MS, Bonner-Weir S, Engelse MA, Rabelink TJ, Heimberg H, Vries RGJ, van Oudenaarden A, Carlotti F, Clevers H, de Koning EJP (2018) Expansion of adult human pancreatic tissue yields organoids harboring progenitor cells with endocrine differentiation potential. Stem Cell Rep 10(3):712–724
López-Otín C, Blasco MA, Partridge L, Serrano M, Kroemer G (2013) The hallmarks of aging. Cell 153(6):1194–1217
Mameishvili E, Serafimidis I, Iwaszkiewicz S, Lesche M, Reinhardt S, Bölicke N, Büttner M, Stellas D, Papadimitropoulou A, Szabolcs M, Anastassiadis K, Dahl A, Theis F, Efstratiadis A, Gavalas A (2019) Aldh1b1 expression defines progenitor cells in the adult pancreas and is required for Kras-induced pancreatic cancer. Proc Natl Acad Sci USA 116(41):20679–20688
McCarthy N, Kraiczy J, Shivdasani RA (2020) Cellular and molecular architecture of the intestinal stem cell niche. Nat Cell Biol 22(9):1033–1041
Menge BA, Tannapfel A, Belyaev O, Drescher R, Müller C, Uhl W, Schmidt WE, Meier JJ (2008) Partial pancreatectomy in adult humans does not provoke beta-cell regeneration. Diabetes 57(1):142–149
Miyazaki S, Tashiro F, Miyazaki J (2016) Transgenic expression of a single transcription factor Pdx1 induces transdifferentiation of pancreatic acinar cells to endocrine cells in adult mice. PLoS One 11(8):e0161190
Muraro MJ, Dharmadhikari G, Grün D, Groen N, Dielen T, Jansen E, van Gurp L, Engelse MA, Carlotti F, de Koning EJ, van Oudenaarden A (2016) A single-cell transcriptome atlas of the human pancreas. Cell Syst 3:385
Nir T, Melton DA, Dor Y (2007) Recovery from diabetes in mice by beta cell regeneration. J Clin Invest 117(9):2553–2561
Otonkoski T, Andersson S, Knip M, Simell O (1988) Maturation of insulin response to glucose during human fetal and neonatal development. Studies with perifusion of pancreatic isletlike cell clusters. Diabetes 37(3):286–291
Otonkoski T, Knip M, Wong I, Simell O (1991) Lack of glucose-induced functional maturation during long-term culture of human fetal islet cells. Life Sci 48(22):2157–2163

Pan FC, Wright C (2011) Pancreas organogenesis: from bud to plexus to gland. Dev Dyn 240(3):530–565

Pan FC, Bankaitis ED, Boyer D, Xu X, Van de Casteele M, Magnuson MA, Heimberg H, Wright CV (2013) Spatiotemporal patterns of multipotentiality in Ptf1a-expressing cells during pancreas organogenesis and injury-induced facultative restoration. Development 140(4):751–764

Papizan JB, Singer RA, Tschen SI, Dhawan S, Friel JM, Hipkens SB, Magnuson MA, Bhushan A, Sussel L (2011) Nkx2.2 repressor complex regulates islet β-cell specification and prevents β-to-α-cell reprogramming. Genes Dev 25(21):2291–2305

Parsons JA, Brelje TC, Sorenson RL (1992) Adaptation of islets of Langerhans to pregnancy: increased islet cell proliferation and insulin secretion correlates with the onset of placental lactogen secretion. Endocrinology 130(3):1459–1466

Perez-Frances M, van Gurp L, Abate MV, Cigliola V, Furuyama K, Bru-Tari E, Oropeza D, Carreaux T, Fujitani Y, Thorel F, Herrera PL (2021) Pancreatic Ppy-expressing γ-cells display mixed phenotypic traits and the adaptive plasticity to engage insulin production. Nat Commun 12(1):4458

Peshavaria M, Larmie BL, Lausier J, Satish B, Habibovic A, Roskens V, Larock K, Everill B, Leahy JL, Jetton TL (2006) Regulation of pancreatic beta-cell regeneration in the normoglycemic 60% partial-pancreatectomy mouse. Diabetes 55(12):3289–3298

Porat S, Weinberg-Corem N, Tornovsky-Babaey S, Schyr-Ben-Haroush R, Hija A, Stolovich-Rain M, Dadon D, Granot Z, Ben-Hur V, White P, Girard CA, Karni R, Kaestner KH, Ashcroft FM, Magnuson MA, Saada A, Grimsby J, Glaser B, Dor Y (2011) Control of pancreatic β cell regeneration by glucose metabolism. Cell Metab 13(4):440–449

Puri S, Roy N, Russ HA, Leonhardt L, French EK, Roy R, Bengtsson H, Scott DK, Stewart AF, Hebrok M (2018) Replication confers β cell immaturity. Nat Commun 9(1):485

Qadir MMF, Álvarez-Cubela S, Klein D, van Dijk J, Muñiz-Anquela R, Moreno-Hernández YB, Lanzoni G, Sadiq S, Navarro-Rubio B, García MT, Díaz Á, Johnson K, Sant D, Ricordi C, Griswold A, Pastori RL, Domínguez-Bendala J (2020) Single-cell resolution analysis of the human pancreatic ductal progenitor cell niche. Proc Natl Acad Sci USA 117(20):10876–10887

Qiu WL, Zhang YW, Feng Y, Li LC, Yang L, Xu CR (2017) Deciphering pancreatic islet β cell and α cell maturation pathways and characteristic features at the single-cell level. Cell Metab 25(5):1194–1205.e1194

Ramzy A, Thompson DM, Ward-Hartstonge KA, Ivison S, Cook L, Garcia RV, Loyal J, Kim PTW, Warnock GL, Levings MK, Kieffer TJ (2021) Implanted pluripotent stem-cell-derived pancreatic endoderm cells secrete glucose-responsive C-peptide in patients with type 1 diabetes. Cell Stem Cell 28(12):2047–2061.e2045

Rankin MM, Kushner JA (2009) Adaptive beta-cell proliferation is severely restricted with advanced age. Diabetes 58(6):1365–1372

Rezanejad H, Ouziel-Yahalom L, Keyzer CA, Sullivan BA, Hollister-Lock J, Li WC, Guo L, Deng S, Lei J, Markmann J, Bonner-Weir S (2018) Heterogeneity of SOX9 and HNF1β in pancreatic ducts is dynamic. Stem Cell Rep 10(3):725–738

Rezanejad H, Lock JH, Sullivan BA, Bonner-Weir S (2019) Generation of pancreatic ductal organoids and whole-mount immunostaining of intact organoids. Curr Protoc Cell Biol 83(1):e82

Rieck S, Kaestner KH (2010) Expansion of beta-cell mass in response to pregnancy. Trends Endocrinol Metab 21(3):151–158

Rooman I, Bouwens L (2004) Combined gastrin and epidermal growth factor treatment induces islet regeneration and restores normoglycaemia in C57Bl6/J mice treated with alloxan. Diabetologia 47(2):259–265

Rooman I, Heremans Y, Heimberg H, Bouwens L (2000) Modulation of rat pancreatic acinoductal transdifferentiation and expression of PDX-1 in vitro. Diabetologia 43(7):907–914

Rorsman P, Ashcroft FM (2018) Pancreatic β-cell electrical activity and insulin secretion: of mice and men. Physiol Rev 98(1):117–214

Rosenberg L (1995) In vivo cell transformation: neogenesis of beta cells from pancreatic ductal cells. Cell Transplant 4(4):371–383

Rovira M, Scott SG, Liss AS, Jensen J, Thayer SP, Leach SD (2010) Isolation and characterization of centroacinar/terminal ductal progenitor cells in adult mouse pancreas. Proc Natl Acad Sci USA 107(1):75–80
Sancho R, Gruber R, Gu G, Behrens A (2014) Loss of Fbw7 reprograms adult pancreatic ductal cells into α, δ, and β cells. Cell Stem Cell 15(2):139–153
Sander M, Sussel L, Conners J, Scheel D, Kalamaras J, Dela Cruz F, Schwitzgebel V, Hayes-Jordan A, German M (2000) Homeobox gene Nkx6.1 lies downstream of Nkx2.2 in the major pathway of beta-cell formation in the pancreas. Development 127(24):5533–5540
Schaffer AE, Yang AJ, Thorel F, Herrera PL, Sander M (2011) Transgenic overexpression of the transcription factor Nkx6.1 in β-cells of mice does not increase β-cell proliferation, β-cell mass, or improve glucose clearance. Mol Endocrinol 25(11):1904–1914
Schaffer AE, Taylor BL, Benthuysen JR, Liu J, Thorel F, Yuan W, Jiao Y, Kaestner KH, Herrera PL, Magnuson MA, May CL, Sander M (2013) Nkx6.1 controls a gene regulatory network required for establishing and maintaining pancreatic Beta cell identity. PLoS Genet 9(1):e1003274
Seymour PA (2014) Sox9: a master regulator of the pancreatic program. Rev Diabet Stud 11(1):51–83
Shapiro AMJ, Thompson D, Donner TW, Bellin MD, Hsueh W, Pettus J, Wilensky J, Daniels M, Wang RM, Brandon EP, Jaiman MS, Kroon EJ, D'Amour KA, Foyt HL (2021) Insulin expression and C-peptide in type 1 diabetes subjects implanted with stem cell-derived pancreatic endoderm cells in an encapsulation device. Cell Rep Med 2(12):100466
Shin JS, Kim JM, Min BH, Chung H, Park CG (2019) Absence of spontaneous regeneration of endogenous pancreatic beta-cells after chemical-induced diabetes and no effect of GABA on alpha-to-beta cell transdifferentiation in rhesus monkeys. Biochem Biophys Res Commun 508(4):1056–1061
Solar M, Cardalda C, Houbracken I, Martín M, Maestro MA, De Medts N, Xu X, Grau V, Heimberg H, Bouwens L, Ferrer J (2009) Pancreatic exocrine duct cells give rise to insulin-producing beta cells during embryogenesis but not after birth. Dev Cell 17(6):849–860
Sorenson RL, Brelje TC (1997) Adaptation of islets of Langerhans to pregnancy: beta-cell growth, enhanced insulin secretion and the role of lactogenic hormones. Horm Metab Res 29(6):301–307
Spears E, Serafimidis I, Powers AC, Gavalas A (2021) Debates in pancreatic beta cell biology: proliferation versus progenitor differentiation and transdifferentiation in restoring β cell mass. Front Endocrinol 12:722250
Stewart AF, Hussain MA, García-Ocaña A, Vasavada RC, Bhushan A, Bernal-Mizrachi E, Kulkarni RN (2015) Human β-cell proliferation and intracellular signaling: part 3. Diabetes 64(6):1872–1885
Stoffers DA, Ferrer J, Clarke WL, Habener JF (1997) Early-onset type-II diabetes mellitus (MODY4) linked to IPF1. Nat Genet 17(2):138–139
Stolovich-Rain M, Enk J, Vikesa J, Nielsen FC, Saada A, Glaser B, Dor Y (2015) Weaning triggers a maturation step of pancreatic β cells. Dev Cell 32(5):535–545
Sussel L, Kalamaras J, Hartigan-O'Connor DJ, Meneses JJ, Pedersen RA, Rubenstein JL, German MS (1998) Mice lacking the homeodomain transcription factor Nkx2.2 have diabetes due to arrested differentiation of pancreatic beta cells. Development 125(12):2213–2221
Takahashi M, Miyatsuka T, Suzuki L, Osonoi S, Himuro M, Miura M, Katahira T, Wakabayashi Y, Fukunaka A, Nishida Y, Fujitani Y, Takeda S, Mizukami H, Itakura A, Watada H (2020) Biphasic changes in β-cell mass around parturition are accompanied by increased serotonin production. Sci Rep 10(1):4962
Talchai C, Xuan S, Lin HV, Sussel L, Accili D (2012) Pancreatic β cell dedifferentiation as a mechanism of diabetic β cell failure. Cell 150(6):1223–1234
Tan SY, Mei Wong JL, Sim YJ, Wong SS, Mohamed Elhassan SA, Tan SH, Ling Lim GP, Rong Tay NW, Annan NC, Bhattamisra SK, Candasamy M (2019) Type 1 and 2 diabetes mellitus: a review on current treatment approach and gene therapy as potential intervention. Diabetes Metab Syndr 13(1):364–372

Tatsuoka H, Sakamoto S, Yabe D, Kabai R, Kato U, Okumura T, Botagarova A, Tokumoto S, Usui R, Ogura M, Nagashima K, Mukai E, Fujitani Y, Watanabe A, Inagaki N (2020) Single-cell transcriptome analysis dissects the replicating process of pancreatic beta cells in partial pancreatectomy model. iScience 23(12):101774

Teta M, Rankin MM, Long SY, Stein GM, Kushner JA (2007) Growth and regeneration of adult beta cells does not involve specialized progenitors. Dev Cell 12(5):817–826

Thorel F, Népote V, Avril I, Kohno K, Desgraz R, Chera S, Herrera PL (2010) Conversion of adult pancreatic alpha-cells to beta-cells after extreme beta-cell loss. Nature 464(7292):1149–1154

Tosti L, Hang Y, Debnath O, Tiesmeyer S, Trefzer T, Steiger K, Ten FW, Lukassen S, Ballke S, Kühl AA, Spieckermann S, Bottino R, Ishaque N, Weichert W, Kim SK, Eils R, Conrad C (2020) Single-nucleus and in situ RNA-sequencing studies reveal cell topographies in the human pancreas. Gastroenterology 160(4):1330–1344

Townsend AR, Skehel JJ (1984) The influenza A virus nucleoprotein gene controls the induction of both subtype specific and cross-reactive cytotoxic T cells. J Exp Med 160(2):552–563

Tschen SI, Dhawan S, Gurlo T, Bhushan A (2009) Age-dependent decline in beta-cell proliferation restricts the capacity of beta-cell regeneration in mice. Diabetes 58(6):1312–1320

van der Meulen T, Mawla AM, DiGruccio MR, Adams MW, Nies V, Dólleman S, Liu S, Ackermann AM, Cáceres E, Hunter AE, Kaestner KH, Donaldson CJ, Huising MO (2017) Virgin beta cells persist throughout life at a neogenic niche within pancreatic islets. Cell Metab 25(4):911–926.e916

Vieira A, Vergoni B, Courtney M, Druelle N, Gjernes E, Hadzic B, Avolio F, Napolitano T, Navarro Sanz S, Mansouri A, Collombat P (2018) Neurog3 misexpression unravels mouse pancreatic ductal cell plasticity. PLoS One 13(8):e0201536

Wang YJ, Kaestner KH (2019) Single-cell RNA-seq of the pancreatic islets—a promise not yet fulfilled? Cell Metab 29(3):539–544

Wang Z, York NW, Nichols CG, Remedi MS (2014) Pancreatic β cell dedifferentiation in diabetes and redifferentiation following insulin therapy. Cell Metab 19(5):872–882

Wang D, Wang J, Bai L, Pan H, Feng H, Clevers H, Zeng YA (2020) Long-term expansion of pancreatic islet organoids from resident procr(+) progenitors. Cell 180(6):1198–1211.e1119

Wang J, Wang D, Chen X, Yuan S, Bai L, Liu C, Zeng YA (2022) Isolation of mouse pancreatic islet Procr. Nat Protoc 17(5):1359–1384

Xiao X, Guo P, Shiota C, Zhang T, Coudriet GM, Fischbach S, Prasadan K, Fusco J, Ramachandran S, Witkowski P, Piganelli JD, Gittes GK (2018) Endogenous reprogramming of alpha cells into beta cells, induced by viral gene therapy, reverses autoimmune diabetes. Cell Stem Cell 22(1):78–90.e74

Xu G, Stoffers DA, Habener JF, Bonner-Weir S (1999) Exendin-4 stimulates both beta-cell replication and neogenesis, resulting in increased beta-cell mass and improved glucose tolerance in diabetic rats. Diabetes 48(12):2270–2276

Xu X, D'Hoker J, Stangé G, Bonné S, De Leu N, Xiao X, Van de Casteele M, Mellitzer G, Ling Z, Pipeleers D, Bouwens L, Scharfmann R, Gradwohl G, Heimberg H (2008) Beta cells can be generated from endogenous progenitors in injured adult mouse pancreas. Cell 132(2):197–207

Yi Z, Waseem Ghani M, Ghani H, Jiang W, Waseem Birmani M, Ye L, Bin L, Cun LG, Lilong A, Mei X (2020) Gimmicks of gamma-aminobutyric acid (GABA) in pancreatic beta-cell regeneration through transdifferentiation of pancreatic alpha- to beta-cells. Cell Biol Int 44(4):926–936

Yoneda S, Uno S, Iwahashi H, Fujita Y, Yoshikawa A, Kozawa J, Okita K, Takiuchi D, Eguchi H, Nagano H, Imagawa A, Shimomura I (2013) Predominance of β-cell neogenesis rather than replication in humans with an impaired glucose tolerance and newly diagnosed diabetes. J Clin Endocrinol Metab 98(5):2053–2061

Zhang M, Lin Q, Qi T, Wang T, Chen CC, Riggs AD, Zeng D (2016) Growth factors and medium hyperglycemia induce Sox9+ ductal cell differentiation into β cells in mice with reversal of diabetes. Proc Natl Acad Sci USA 113(3):650–655

Zhao H, Huang X, Liu Z, Pu W, Lv Z, He L, Li Y, Zhou Q, Lui KO, Zhou B (2021) Pre-existing beta cells but not progenitors contribute to new beta cells in the adult pancreas. Nat Metab 3(3):352–365

Zhou Q, Law AC, Rajagopal J, Anderson WJ, Gray PA, Melton DA (2007) A multipotent progenitor domain guides pancreatic organogenesis. Dev Cell 13(1):103–114

Zhou Q, Brown J, Kanarek A, Rajagopal J, Melton DA (2008) In vivo reprogramming of adult pancreatic exocrine cells to beta-cells. Nature 455(7213):627–632

Metabolic and Molecular Amplification of Insulin Secretion

Mourad Ferdaoussi

Abstract The pancreatic β cells are at the hub of myriad signals to regulate the secretion of an adequate amount of insulin needed to re-establish postprandial euglycemia. The β cell possesses sophisticated metabolic enzymes and a variety of extracellular receptors and channels that amplify insulin secretion in response to autocrine, paracrine, and neurohormonal signals. Considerable research has been undertaken to decipher the mechanisms regulating insulin secretion. While the triggering pathway induced by glucose is needed to initiate the exocytosis process, multiple other stimuli modulate the insulin secretion response. This chapter will discuss the recent advances in understanding the role of the diverse glucose- and fatty acid-metabolic coupling factors in amplifying insulin secretion. It will also highlight the intracellular events linking the extracellular receptors and channels to insulin secretion amplification. Understanding these mechanisms provides new insights into learning more about the etiology of β-cell failure and paves the way for developing new therapeutic strategies for type 2 diabetes.

Keywords Amplifying pathway · Insulin secretion · Islets · Neurohormones · Fatty acids · NADPH · Incretins

Abbreviations

6GPDH	6-Phosphogluconate dehydrogenase
ABHD6	α/β-Hydrolase domain containing 6
AC	Adenylyl cyclase
ACC1	Acyl-CoA carboxylase 1

M. Ferdaoussi (✉)
Faculty Saint-Jean and Alberta Diabetes Institute, University of Alberta, Edmonton, AB, Canada
e-mail: Mourad.Ferdaoussi@ualberta.ca

M. L. Golson (ed.), *Molecular and Cell Biology of Pancreas Development, Function and Regeneration*, Advances in Anatomy, Embryology and Cell Biology 239, https://doi.org/10.1007/978-3-031-62232-8_5

ACLY	ATP citrate lyase
ACO1	Aconitase 1
CGI58	Comparative gene identification-58
CIC	Citrate-isocitrate carrier
CPT1	Carnitine palmitoyltransferase I
DAG	Diacylglycerols
EPAC	Exchange protein activated by cAMP
FFA	Free fatty acids
FFAR	FFA receptor
G3P	Glycerol-3 phosphate
G3PP	Glycerol-3-phosphate phosphatase
G6PDH	Glucose-6-phosphate dehydrogenase
GIP	Glucose-dependent insulinotropic polypeptide
GLP-1	Glucagon-like peptide-1
GLUT	Glucose transporter
GRX1	Glutaredoxin 1
GSH	Glutathione
GSIS	Glucose-stimulated insulin secretion
GSSG	Glutathione disulfide
IDH	Isocitrate dehydrogenase
IP3	Inositol trisphosphate
M3R	M3 muscarinic receptor
MAG	Monoacylglycerols
MDH1	NAD^+-dependent isocitrate dehydrogenase
MEc	Cytosolic malic enzyme
PAK4	p21 activated kinase 4
PC	Pyruvate carboxylase
PDH	Pyruvate dehydrogenase
PEP	Pannexin-1 (Panx1) phosphoenolpyruvate
PEPCK	Phosphoenolpyruvate carboxykinase
PK	Pyruvate kinase
PKD1	Protein kinase D1
PLC	Phospholipase C
SCS-ATP	Succinyl-CoA synthetase ATP
SCS-GTP	Succinyl-CoA synthetase GTP
SENP1	Sentrin-specific protease 1
SNARE	Soluble *N*-ethylmaleimide-sensitive factor attachment protein receptor
T1D	Type 1 diabetes
T2D	Type 2 diabetes
TCA	Citric acid cycle
TG	Triacylglycerol
TR	Sweet taste receptors
TRPC3	Transient receptor potential channel 3
TRPC3	Transient receptor potential channel 3
VDCC	Voltage-gated calcium channel

A combination of metabolic and neurohormonal signals tightly regulates insulin secretion from pancreatic β cells to ensure adequate regulation of postprandial glucose concentrations. Secreted insulin facilitates the storage of glucose or its cellular usage. Inadequate insulin secretion alters glucose homeostasis. The autoimmune destruction of β cells causes type 1 diabetes (T1D) (DiMeglio et al. 2018), whereas the decline of β-cell secretory capacity and β-cell death are central to type 2 diabetes (T2D) etiology (Prentki and Nolan 2006). A malfunction in β-cell regulation, usually accompanied by an expansion of β-cell mass, results in congenital hyperinsulinism, a disease characterized by hyperinsulinemic hypoglycemia (Sims 2021). Thus, the pancreatic β cells use an arsenal of molecular mechanisms to couple the sensing of circulating glucose and other nutrients to release the right amount of insulin to re-establish euglycemia.

The cellular machinery of β cells rapidly detects the increase of postprandial glucose and adjusts the insulin secretion response to re-establish homeostatic conditions. Glucose is the primary physiological insulin secretagogue required to prime the exocytotic machinery to stimulate insulin secretion biphasically. The first, short phase occurring during the first 5–10 min is recognized as the triggering pathway, followed by a sustained release of insulin, which corresponds to the amplifying pathway (Henquin 2000). The cellular events of the triggering pathway involve the entry of glucose via glucose transporters (GLUT1 and GLUT2 in humans, and GLUT2 in rodents) and its metabolism by the mitochondrial citric acid cycle (TCA), which increases the intracellular ATP/ADP ratio that causes ATP-sensitive K^+ (K_{ATP}) channel closure, and thus leading to membrane depolarization and the opening of the voltage-gated calcium channel (VDCC) (Fig. 1). Elevation of the intracellular calcium concentration regulates soluble *N*-ethylmaleimide-sensitive factor attachment protein receptor (SNARE) proteins of the exocytotic machinery to achieve the transport and the fusion of insulin-containing granules with the plasma membrane to release insulin into the bloodstream. However, this simplified model was challenged by recent studies showing a spatial and temporal plasma membrane compartmentalization of differing ATP/ADP ratios (reviewed Merrins et al. 2022). On the other hand, the amplifying pathway controls the amplitude of secreted insulin to match the body's needs (Ferdaoussi and MacDonald 2017). This chapter will focus on the mechanisms regulating the amplification of insulin secretion by glucose and other nutrients, as well as by neurohormonal signals.

1 The Role of Glucose-Metabolic Coupling Factors on Insulin Secretion Amplification

The first evidence of the amplification of insulin secretion by glucose independent of its role in K_{ATP} channel closure was suggested by early studies by J.C Henquin (Gembal et al. 1993) and T. Yamada (Sato et al. 1992). They blocked glucose-stimulated insulin secretion (GSIS) with the K_{ATP} channel opener diazoxide and

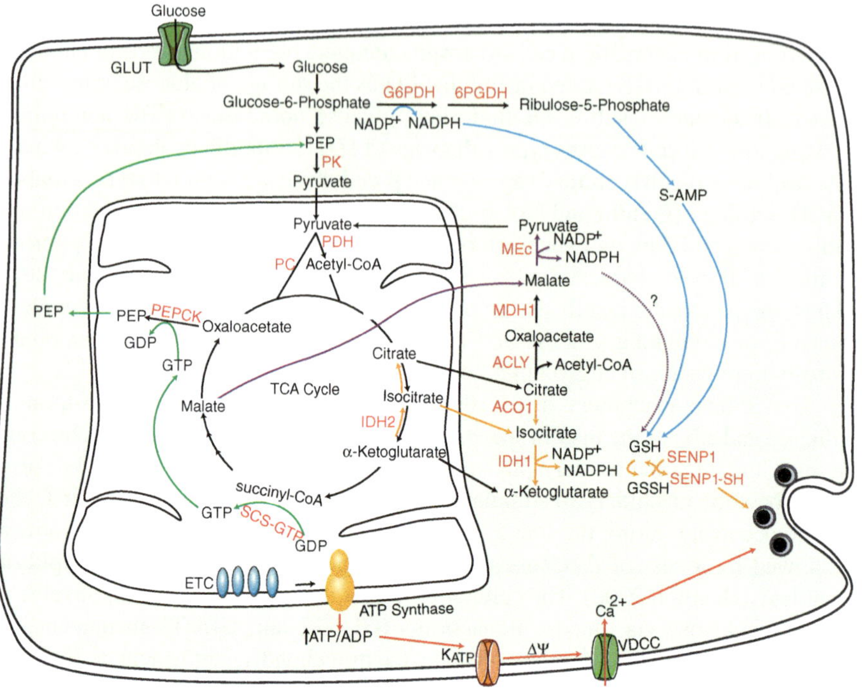

Fig. 1 Metabolic amplification of insulin secretion by glucose metabolism. Intracellular glucose metabolism produces multiple coupling factors to amplify insulin secretion. These coupling factors are generated by the citrate-isocitrate pathway (orange arrows), pentose pathway (blue arrows), malic enzyme pathway (purple arrows), and phosphoenolpyruvate pathway (green arrows). The stimulation of the triggering pathway (red arrows) is required to prime the insulin secretion process. The key enzyme regulating the amplification of insulin secretion are 6PGDH, 6-phosphogluconate dehydrogenase; ACLY, ATP citrate lyase; ACO1, aconitase 1; ETC, electron transport chain; G6PDH, glucose-6-phosphate dehydrogenase; GLUT, glucose transporters; IDH1 and IDH2, isocitrate dehydrogenase 1 and 2; KATP, ATP-sensitive K+ channel; PC, pyruvate carboxylase; PDH, pyruvate dehydrogenase; PEPCK, phosphoenolpyruvate carboxykinase; PK, pyruvate kinase; MDH1, NAD+-dependent isocitrate dehydrogenase; MEc, cytosolic malic enzyme; SCS-GTP, succinyl-CoA synthetase GTP; SENP1, sentrin-specific protease 1; VDCC, voltage-gated calcium channel

generated an independent membrane depolarization with KCl. In this experimental setting, glucose-induced insulin secretion in a K_{ATP}-independent manner. This mechanism is currently recognized as the amplifying pathway and can also be measured at the single-cell level. Indeed, the exocytotic response, measured by patch clamp, was enhanced when human β cells were incubated in high glucose (10 mM) compared to low glucose (1 mM) (Ferdaoussi et al. 2015). Interestingly, the amplifying pathway was strongly impaired in β cells from T2D donors (Ferdaoussi et al. 2015).

The carbons within glucose are mostly oxidized in the mitochondria (Schuit et al. 1997) to produce numerous metabolic coupling factors that regulate intracellular pathways and facilitate insulin-containing granules exocytosis. Upon its entry, glucose is phosphorylated to Glucose-6-phosphate and then metabolized to pyruvate. The latter is used by pyruvate dehydrogenase (PDH) or pyruvate carboxylase (PC) to generate acetyl-CoA and oxaloacetate, respectively (Schuit et al. 1997) (Fig. 1). The challenging question that attracted the focus of multidisciplinary research is which factor(s) couple(s) glucose metabolism with insulin secretion.

1.1 Cytosolic NADPH

A high glucose concentration enhances the TCA cycle and the anaplerotic flux to generate citrate, isocitrate, α-ketoglutarate or malate that can be shuttled between the mitochondria and cytosol to contribute to cytosolic NADPH production (Ishihara 2022) (Fig. 1). Intracellular NADPH levels correlate positively with glucose concentration (Ivarsson et al. 2005) and insulin secretion in the pancreatic β cell line INS 832/13 (Ronnebaum et al. 2006). Different pathways generate cytosolic NADPH from the factors shuttling between the mitochondria and cytosol. First, the cytosolic isocitrate dehydrogenase 1 (IDH1) reduces $NADP^+$ to NADPH to catalyze α-ketoglutarate production from isocitrate (Joseph et al. 2006; Guay et al. 2013) (Fig. 1). Second, cytosolic malic enzyme (MEc) converts malate to pyruvate by producing NADPH (Ronnebaum et al. 2008) (Fig. 1). Third, Aconitase 1 (ACO1) can indirectly elevate cytosolic NADPH levels by converting citrate to isocitrate, thereby increasing available substrate for IDH1 (Fig. 1). Fourth, ATP citrate lyase (ACLY) and NAD^+-dependent isocitrate dehydrogenase (MDH1) can also indirectly elevate cytosolic NADPH levels by converting citrate to oxaloacetate and then malate, respectively. Malate is then used by MEc to enhance cytosolic NADPH levels (Azzouny et al. 2016) (Fig. 1).

Direct infusion of NADPH into patch-clamped healthy human β cells mimics the effect of high glucose concentration on amplifying the exocytosis elicited by electrical depolarization (Ferdaoussi et al. 2015; Ivarsson et al. 2005). Interestingly, none of the other potential metabolites derived from glucose oxidation, such as α-ketoglutarate, phosphoenolpyruvate (PEP), or NADH, were able to amplify exocytosis (Ferdaoussi et al. 2015). An infusion of isocitrate provides more substrate for IDH1 and increases the cytosolic levels of β cells NADPH, resulting in amplified exocytosis. Inversely, siRNA knockdown of IDH1 in human β cells or INS 832/13 cells reduces exocytosis and insulin secretion (Ferdaoussi et al. 2015). Interestingly, the reintroduction of NADPH but not isocitrate restores the loss of exocytotic amplification mediated by IDH1 knockdown. These molecular manipulations (Ferdaoussi et al. 2015) suggest that NADPH produced by IDH1 is a potent glucose-coupling factor in insulin secretion.

A recent study (Zhang et al. 2021) showed that mitochondrial isocitrate dehydrogenase (IDH2), an enzyme that regulates the reductive TCA cycle (i.e., counterclockwise) increases mitochondrial isocitrate levels by reducing α-ketoglutarate using $NADP^+$ as a cofactor (Zhang et al. 2021) (Fig. 1). The mitochondrial isocitrate or its isomer citrate are then transported to the cytosol, where they enhance cytosolic NADPH levels. The knockdown (Zhang et al. 2021; Macdonald et al. 2013) or the inhibition (Zhang et al. 2021) of IDH2 in INS 832/13 cells and rodent β cells reduces cytosolic NADPH and insulin secretion. The discovery of the reductive TCA cycle in β cells added more comprehension of the metabolic mechanisms regulating insulin secretion and support the role of cytosolic NADPH as a glucose-coupling factor in insulin secretion.

The oxidation of malate to pyruvate by MEc enhances cytosolic NADPH levels. Various groups have reported conflicting observations on the role of MEc in INS 832/13 cells or rodent islets (Xu et al. 2008; Guay et al. 2007; Pongratz et al. 2007). However, islets isolated from MEc null mice or rodent islets that underwent siRNA knockdown of MEc show no alterations in GSIS (Ronnebaum et al. 2008), suggesting that MEc-dependent production of NADPH is unlikely to be necessary for the regulation of insulin secretion.

The downstream mechanism currently proposed to link glucose-induced cytosolic NADPH production to exocytosis involves the reduction of glutathione (GSH) to glutathione disulfide (GSSG) by glutaredoxin 1 (GRX1) (Ferdaoussi et al. 2015). Then, GSSH reduces and activates Sentrin-specific protease 1 (SENP1) (Fig. 1). SENP1 modifies proteins posttranslationally by cleaving conjugated SUMO 1–3 peptides from SUMOylated proteins (Jansen and Vertegaal 2021). The exocytotic proteins synaptotagmin VII, syntaxin-1, RIM1α, tomosyn-1, and secretagogin (Ferdaoussi et al. 2017; Craig et al. 2015; Dai et al. 2011; Girach et al. 2013) or the potassium channel Kv2.1 (Dai et al. 2009) were shown to be regulated by SUMOylation/DeSUMOylation and proposed as the downstream targets linking NADPH > GSH > SENP1 to exocytosis. β cell- or islet-specific SENP1 knockout mice are glucose intolerant with reduced in vivo insulin levels (Ferdaoussi et al. 2015; Ferdaoussi et al. 2020), a phenotype aggravated in response to a high-fat diet (Lin et al. 2021). The ex vivo insulin secretion and exocytosis measured in single β cells isolated from these mice or from human β cells transfected with SENP1 siRNA was lowered (Ferdaoussi et al. 2015, 2020), presumably due to a reduction in the glucose-induced amplifying pathway. Additionally, the inhibition of IDH2 reduced NADPH production and increased protein SUMOylation, likely because of reduced SENP1 activity (Ferdaoussi et al. 2015). Remarkably, the overexpression of SENP1 or infusion of its upstream regulators (i.e., NADPH or GSH) restored the impaired amplifying pathway observed in human β cells isolated from T2D donors (Ferdaoussi et al. 2015).

The role of cytosolic NADPH in amplifying insulin secretion is still debated (Merrins et al. 2022; Ferdaoussi and MacDonald 2017; Campbell and Newgard 2021), mainly due to a series of studies. The first demonstrated that the knockdown of IDH1 increases insulin secretion in INS 832/13 cells and rat islets (Guay et al. 2013). A second study showed that the deletion of the gene encoding the citrate-isocitrate carrier (CIC), thereby inhibiting the flux of citrate between the mitochondria and cytosol, significantly reduces NADPH production, without any change in β-cell secretory capacity in response to glucose or other secretagogues (Bauchle et al. 2021). Interestingly, the reduced GSH levels were maintained in β cells isolated from CIC knockout mice, implying that the pathway regulating exocytosis downstream of GSH (Ferdaoussi et al. 2015) is likely preserved in these mice. This result raises the question of whether the production of other reducing-coupling factors might compensate for the absence of CIC and NADPH depletions. Thus, additional targeted experiments are still needed to clarify the role of cytosolic NADPH as a glucose-coupling factor in insulin secretion.

1.2 Pentose Pathway Coupling Factors

The pentose phosphate metabolic pathway produces NADPH by metabolizing glucose-6-phosphate to ribulose-5-phosphate by glucose-6-phosphate dehydrogenase (G6PDH) and 6-phosphogluconate dehydrogenase (6GPDH) (Spégel et al. 2013; Goehring et al. 2011) (Fig. 1). Inhibition or knockdown of these two key enzymes decreases GSIS (Spégel et al. 2013; Goehring et al. 2011). A reduction of the NADPH levels could explain this reduction in insulin secretion (Ferdaoussi et al. 2015; Spégel et al. 2013). Alternatively, de novo purine synthesis, particularly adenylosuccinate (S-AMP), is a coupling factor that mediates insulin secretion amplification by the pentose phosphate metabolic pathway (Gooding et al. 2015) (Fig. 1). Indeed, the inhibition or the knockdown of adenylosuccinate synthase, an enzyme producing the S-AMP from aspartic acid or inosine monophosphate, impaired insulin secretion and exocytosis in islets and INS 832/13 cell lines. Remarkably, the amplification of exocytosis by S-AMP in isolated human β cells was lost in the absence of SENP1 (Gooding et al. 2015), suggesting that the pentose pathway regulates deSUMOylation of exocytotic proteins through SENP1 activation by either NADPH or S-AMP.

1.3 Phosphoenolpyruvate

The mitochondrial oxaloacetate produced by the TCA cycle or by anaplerosis can be metabolized to PEP by phosphoenolpyruvate carboxykinase (PEPCK) in a GTP-dependent manner (Lu et al. 2002; Kibbey et al. 2007) (Fig. 1). The GTP required for this reaction is produced by mitochondrial succinyl-CoA synthetase GTP (SCS-GTP) (Fig. 1). Succinyl-CoA synthetase ATP (SCS-ATP) is another SCS isoform that competes with SCS-GTP for triphosphate nucleotide synthesis by producing ATP. PEP is then transported to the cytosol and metabolized to pyruvate by pyruvate kinase (PK) to refill the TCA cycle and enhance the amplification of insulin secretion (Fig. 1). The knockdown of SCS-GTP, but not SCS-ATP, reduces GSIS; conversely, the overexpression of SCS-GTP enhances insulin secretion (Kibbey et al. 2007). The disturbance of mitochondrial and cytosolic calcium homeostasis is another effect of the modulation of SCS-GTP expression levels (Kibbey et al. 2007). Interestingly, the PK activator enhances the PEP cycle and improves glucose homeostasis, at least in part by enhancing insulin secretion; these data provide more support for the role of PEP as a metabolic coupling factor (Abulizi et al. 2020).

2 The Role of Fatty Acids and Fructose on Insulin Secretion Potentiation

Circulating nutrients such as fatty acids and fructose modulate insulin secretion by regulating metabolic- and receptor-coupled pathways (Dobbins et al. 1998). Importantly, these nutrients do not promote insulin secretion without stimulating the triggering pathway (Dobbins et al. 1998); therefore, this regulation is referred to as insulin secretion potentiation.

2.1 Crosstalk Between Glucose and Fatty Acid Metabolisms on Insulin Secretion Potentiation

Citrate generated by the TCA cycle and transported to the cytosol by CIC can be metabolized by ACLY to produce Acetyl-CoA, a precursor of the malonyl-CoA produced by acyl-CoA carboxylase 1 (ACC1) (Prentki et al. 2020) (Fig. 2). Malonyl-CoA plays a critical role as a metabolic switch to favor free fatty acid (FFA) esterification through the glycerolipid/FFA cycle and to reduce Carnitine Palmitoyltransferase I (CPT1) activity and thereby mitochondrial fatty acid transport and oxidation (Roduit et al. 2004; Prentki et al. 1992) (Fig. 2). Inhibition of FFA oxidation by malonyl-CoA (Roduit et al. 2004), synergistically with glycerol-3 phosphate produced in high glucose conditions (Mugabo et al. 2016), enhances FFA esterification and generation of different lipid species, including triacylglycerol (TG), diacylglycerols (DAG), and monoacylglycerols (MAG) (Prentki et al. 2020) (Fig. 2).

Multiple lines of evidence support the role of fatty acid esterification in regulating insulin secretion, such as the glucose intolerance and reduced insulin secretion observed in mice with a β-cell-specific deletion of the gene encoding ACC1 (Cantley et al. 2019). In addition, the inhibition or knockdown of glycerol-3-phosphate phosphatase (G3PP), the enzyme that hydrolyzes glycerol-3 phosphate (G3P), enhances the glycerolipid/FFA cycle and insulin section (Mugabo et al. 2016).

The TG produced by the glycerolipid/FFA cycle is hydrolyzed by comparative gene identification-58 (CGI58) and adipose triglyceride lipase to produce 2,3 DAG and 1,3 DAG, respectively (Brown et al. 2010). Since these two DAG species have low affinity to C1 and C2 DAG-binding domains (Eichmann and Lass 2015), it is unclear how they mediate their effect on insulin secretion potentiation by regulating DAG-binding proteins. One proposed mechanism involves the hydrolysis of 2,3 DAG to 2-MAG by hormone-sensitive lipase (HSL) to increase the pool of FFA for esterification (Zhao et al. 2014). Also, HSL hydrolyzes 1,3 DAG to 1-MAG, which binds and activates the exocytotic protein Munc-13 to facilitate insulin granule release (Zhao et al. 2014) (Fig. 2). The protection of 1-MAG from hydrolysis by the β-cell-specific deletion of the gene encoding α/β-Hydrolase domain containing 6 (ABHD6) enhances insulin secretion (Zhao et al. 2014) and argues in favor of a role for 1-MAG as a glucose-coupling factor to amplify insulin secretion (Zhao et al. 2014).

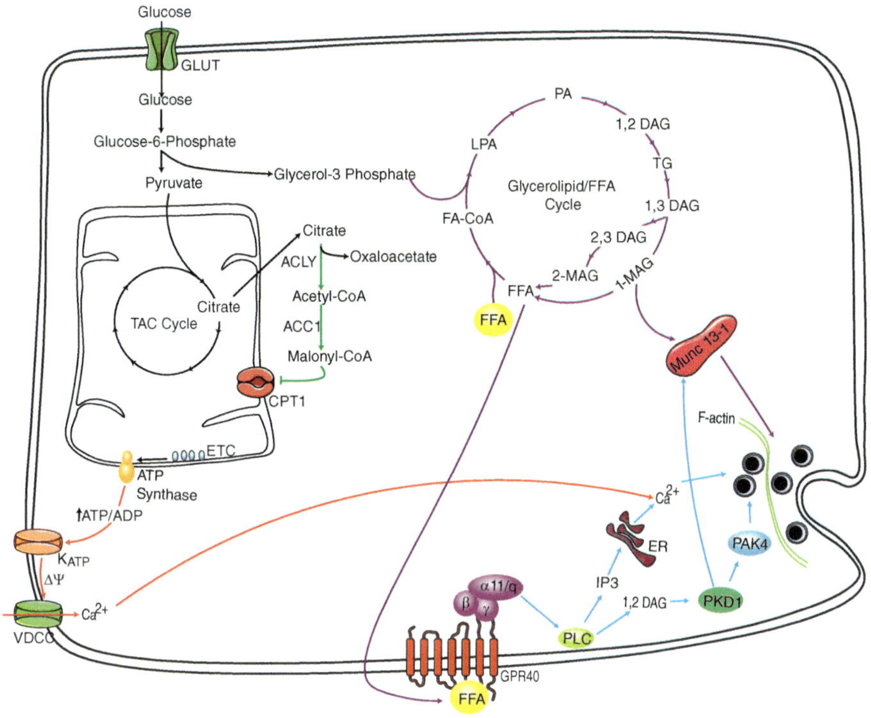

Fig. 2 Fatty acid-dependent potentiation of insulin secretion. Generation of second messengers (DAGs and MAGs) by glycerolipid/FFA cycle (purple arrows) is favored by the inhibition of β oxidation by malonyl-CoA (green arrows), which in synergy with GPR40 pathway activation (blue arrows), potentiates the glucose-induced triggering pathway (red arrows). CPT1, carnitine palmitoyltransferase I; DAG, diacylglycerol; ETC, electron transport chain; ER, endoplasmic reticulum; FA-CoA, fatty acyl-CoA; FFA, free fatty acids; IP3, inositol trisphosphate; KATP, ATP-sensitive K^+ channel; LPA, lysophosphatidic acid; MAG, monoacylglycerol; PAK4, p21 activated kinase 4; PA, phosphatidic acid; PLC, phospholipase C; TG, triacylglycerol, VDCC, voltage-gated calcium channel

2.2 *Potentiation of Insulin Secretion by Fatty Acids Beyond Metabolism*

FFA produced by the glycerol/FFA cycle (Prentki et al. 2020; Hauke et al. 2018) and released by β cells into the extracellular milieu (Martins et al. 2004; Trauelsen et al. 2018), as well as the circulating FFA, are ligands of a heterogeneous G-coupled protein FFA receptor (FFAR) family involved in the regulation of insulin secretion (Ghislain and Poitout 2021; Kebede et al. 2008; Alquier et al. 2009). The nature, length, and degree of FFA saturation determine the specificity and the effect of FFARs on insulin secretion. GPR40 (also known as FFAR1 or FFA1) is abundantly expressed in β cells and is the main FFAR mediating the medium-to-long chain FFA potentiation of GSIS (Latour et al. 2007). GPR40 is also activated by

20-hydroxyeicosatetraenoic acid (20-HETE) produced from arachidonic acid and secreted by β cells (Trauelsen et al. 2018). GPR120 (FFAR4 or FFA4) is also preferably activated by long-chain fatty acids. It indirectly potentiates GSIS by reducing δ-cell somatostatin secretion (Croze et al. 2021). GPR42 (FFAR2, FFA2) and GRP41 (FFAR3 or FFA3) are activated by short-chain fatty acids (Ang et al. 2018), and their direct role in insulin secretion regulation is still debated (Tang et al. 2015; McNelis et al. 2015). This chapter will discuss the mode of action by which FFARs have a direct positive insulinotropic effect on β cells. The role of the other fatty acid receptors was extensively discussed in a previous review (Ghislain and Poitout 2021).

GPR40

GPR40 partially mediates the acute effect of FFA on GSIS potentiation. Indeed, whole-body GPR40 knockout mice have a 60% reduction in insulin secretion during a hyperglycemic clamp (Alquier et al. 2009). Similarly, a single point mutation in the gene encoding GPR40 (R258W) partially impairs insulin secretion in mice fed a high-fat diet, as has also been observed in GPR40 knockout mice (Sabrautzki et al. 2017). Ex vivo, the potentiation of GSIS in response to fatty acids or GPR40 agonist is partially reduced in islets isolated from GPR40 knockout or mice with a GPR40 R258W mutation (Sabrautzki et al. 2017; Ferdaoussi et al. 2012). Orlistat, a panlipase inhibitor, partially reduces the FFA-mediated potentiation of insulin secretion, suggesting that the other effects of fatty acids on GSIS potentiation are mediated by intracellular FFA metabolism through the glycerol/FFA cycle detailed above (Zhao et al. 2014). Remarkably, the deletion of GPR40 does not affect fatty acid metabolism (Alquier et al. 2009), suggesting the absence of the crosstalk between these pathways.

GPR40 is a $G_{\alpha q/11}$-coupled receptor that activates phospholipase C (PLC) in a ligand-dependent manner to produce the second messengers' inositol trisphosphate (IP3) and DAG (particularly 1,2 DAG) from phosphatidylinositol 4,5-bisphosphate (PIP2) (Briscoe et al. 2003). IP3 enhances intracellular calcium levels either from extracellular sources through the L-type Ca^{2+} channel or from the endoplasmic reticulum stores through the IP3R1 receptor (Usui et al. 2019) (Fig. 2). In addition, the PLC activation downstream of GPR40 induces plasma membrane depolarization and exocytosis by stimulating the transient receptor potential channel 3 (TRPC3), a short nonselective cation channel (Yamada et al. 2016). The other arm regulating GSIS potentiation by the GPR40 pathway involves the activation of Protein Kinase D1 (PKD1) by DAG (Ferdaoussi et al. 2012) (Fig. 2). PKD1 then triggers actin filament depolymerization to facilitate the recruitment of insulin granules to exocytotic sites, potentially by regulating the activity of p21 activated kinase 4 (PAK4) (Bergeron et al. 2016), a protein involved in the control of actin filament remodeling (Fig. 2). Notably, β-cell deletion of PKD1 or whole-body GPR40 knockout results in a similar reduction of insulin secretion in mice fed with a high-fat diet (Bergeron et al. 2018). PKD1 activation was abolished entirely in islets isolated from GPR40 knockout mice (Ferdaoussi et al. 2012), implying that the 1-MAG, 1,3 DAG, and 2,3 DAG species produced by the FFA/glycerolipid cycle discussed above are unable to use the PKD1 pathway to amplify insulin secretion.

Munc13-1 (Rhee et al. 2002) is another potential exocytotic protein with two DAG-binding domains that may control exocytosis and GSIS potentiation downstream of GPR40 (Fig. 2).

Because GPR40 potentiates insulin secretion, and GPR40 does not mediate the deleterious effects induced during chronic hyperlipidemia (Kebede et al. 2008; Alquier and Poitout 2009), a GPR40 agonist has a strong potential as a therapeutic drug in patients with T2D. Furthermore, a GPR40 agonist should pose no risk of hypoglycemia since acute GPR40 activation does not stimulate insulin secretion at low glucose concentrations (Ferdaoussi et al. 2012). However, a phase 3 clinical trial of the GPR40 agonist TAK-875 was discontinued due to liver toxicity, challenging this expectancy (Kaku et al. 2015). Nonetheless, a detailed study showed that TAK-875 is a partial GPR40 agonist (Mancini et al. 2015), and unlike fatty acids that potentiate only the second phase of GSIS, TAK-875 potentiates both phases of insulin secretion (Yashiro et al. 2012). Additionally, the liver toxicity could be classdependent since other GPR40 agonists did not show similar adverse effects. Interestingly, in a promising, recent phase 1 clinical trial, SCO-267, a full GPR40 agonist administered orally once daily, stimulated insulin and gut hormone (e.g., GLP-1 and GIP) secretion and improved glycemic control (Nishizaki et al. 2021).

GPR41/GPR43

Short-chain fatty acids are shared ligands for GPR41 and GPR43 (Briscoe et al. 2003). GPR41 is a $G_{\alpha i/o}$-coupled receptor that reduces insulin secretion by decreasing cAMP levels (Fig. 3). However, studies have reported opposing effects upon GPR43 activation (Priyadarshini et al. 2015). While one study showed that GPR43 partially mediates the inhibition of GSIS by acetate (Tang et al. 2015), another study showed that a GPR43 synthetic agnostic potentiates GSIS (McNelis et al. 2015). GPR43 is a dual $G_{\alpha i/o}$- and $G_{\alpha q/11}$-coupled receptor (Tang et al. 2015; McNelis et al. 2015; Priyadarshini et al. 2015) (Fig. 3). While $G_{\alpha i/o}$ activation reduces the cAMP levels and insulin secretion, $G_{\alpha q/11}$ increases IP3 and DAG levels to potentiate GSIS. Thus, these conflicting observations might be due to a ligand-specific stimulation of GPR43 downstream pathways. Investigating the precise role of GPR43 in regulating insulin secretion and glucose homeostasis will provide more insight into whether a GPR43 agonist or antagonist is suitable as a T2D drug.

2.3 Potentiation of Insulin Secretion by Fructose and Artificial Sweeteners

Dietary fructose stimulates sweet taste perception more powerfully than other sugars and is known to potentiate GSIS from human pancreatic islets (Just et al. 2008). While β cells do not express the correct enzymes to metabolize fructose efficiently, the sweet taste receptors (TR) T1R2 and T1R3 are expressed and activated in β cells and mediate the potentiation of GSIS by fructose (Kyriazis et al. 2012) (Fig. 3).

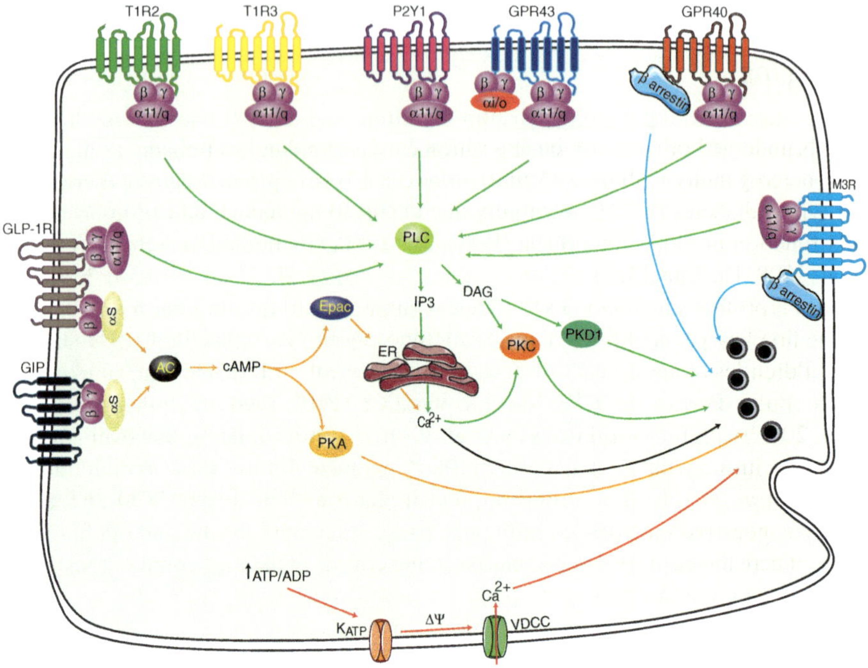

Fig. 3 G-protein coupled receptors-dependent potentiation of GSIS. Generation of second messengers IP3 and DAG, downstream of Gαq/11-coupled receptors (green arrows) and cAMP, downstream of Gαs-coupled receptors (orange arrows), as well as the activation of β arrestin (blue arrow) regulate the activity of downstream targets to potentiate the glucose-stimulated triggering pathway (red arrows). AC, adenylyl cyclase; DAG, diacylglycerol; ER, endoplasmic reticulum; EPAC, exchange protein activated by cAMP; IP3, inositol trisphosphate; GIP, glucose-dependent insulinotropic polypeptide receptor; GLP-1R, glucagon-like peptide-1 receptor; GPR40, G-protein-coupled receptor 40; GPR43, G-protein-coupled receptor 43; M3R, M3 muscarinic receptor; P2Y1, purinergic G protein-coupled receptor P2Y1; T1R2, sweet taste receptor T1R2; T1R3, sweet taste receptors T1R3; PKA, protein kinase A; PKC, protein kinase C; PKD, protein kinase D; VDCC, voltage-gated calcium channel

Saccharin, currently used as a sugar substitute because of its lower caloric intake, activates taste receptors and mimics the effects of fructose on amplifying GSIS (Kyriazis et al. 2012; Serrano et al. 2022). T1R2 and T1R3 are $G_{\alpha q/11}$-coupled receptors that potentiate insulin secretion by producing the second messengers IP3 and DAG (Serrano et al. 2022) (Fig. 3).

3 Neurohormonal Regulation of Insulin Secretion Potentiation

The β cell is at the junction of multiple physiological signals and uses a panoply of signaling pathways to coordinate insulin secretion in response to the body's needs.

3.1 Autocrine Amplification of Insulin Secretion

It has been postulated that the β cell auto-adjusts its function by releasing several factors that act in a positive autocrine feedback loop to amplify insulin secretion. These factors include ATP and glycine, in addition to FFA and 20-HETE described above.

ATP

In β cells, ATP is released through the single-membrane channel Pannexin-1 (Panx1) (Bhat and Sajjad 2021) or co-secreted with insulin (Braun et al. 2007). Cumulative evidence supports a positive insulinotropic effect of ATP on insulin secretion through purinergic receptors. Further, the protection of extracellular ATP from degradation by the ectonucleotidase NTPDase3 enhances GSIS (Syed et al. 2013). Mechanistic studies have shown that incubating human and rodent β cells with ATP increases intracellular calcium levels, plasma membrane depolarization, and insulin secretion (Khan et al. 2014). While one study reported that ATP induced a sharp transient increase of intracellular calcium from the endoplasmic reticulum stores in human β cells (Khan et al. 2014), a second study could not confirm this observation (Jacques-Silva et al. 2010). However, these two studies agreed that in human β cells, ATP increased intracellular calcium concentration from extracellular sources.

P1 and P2 are the two classes of purinergic receptors activated by ATP (Burnstock 2018). P1 receptors belonging to the G-protein-coupled receptor family do not mediate the effect of ATP on insulin secretion (Jacques-Silva et al. 2010). The P2 receptors are divided into two subclasses. The ligand-gated ion channel P2X receptors (P2X 1–7) are permeable to calcium, potassium, and sodium. The P2Y receptors (P2Y 1, 2, 4, 6, 11–14) are G-protein-coupled receptors (Burnstock 2018). The role of each subclass of purinergic receptors on insulin secretion is still unclear and may vary by species. Indeed, it has been proposed that P2X3 in human β cells (Jacques-Silva et al. 2010) or P2X7 (Tozzi et al. 2018) in rodent β cells increases the late flux of Na^+ and Ca^{2+}, resulting in membrane depolarization and the rise of calcium entry through the high-voltage-gated Ca^{2+} channels and thereby leading to exocytosis amplification. However, other studies showed that a P2X (1–3) agonist has little effect on intracellular calcium concentration and exocytosis in human β cells, whereas a P2Y1 inhibitor impaired exocytosis and GSIS (Khan et al. 2014, 2019; Wuttke et al. 2013; Léon et al. 2005). P2Y1 is a $G_{\alpha q/11}$-coupled receptor that generates the second messengers IP3 and DAG upon activation. DAG downstream of P2Y1 activates the protein kinase C (PKC) or PKD1 (Khan et al. 2019) (Fig. 3). However, the increase of GSIS from P2Y1-knockout islets adds more confusion about the role of this receptor on insulin secretion (Léon et al. 2005). Therefore, although the role of extracellular ATP as an autocrine insulin secretagogue is demonstrated in human and rodent β cells, additional investigations are still needed to provide insight into the mechanism by which ATP amplifies insulin secretion.

Glycine
Glycine is a non-essential amino acid that can act as an autocrine signal to amplify insulin secretion. The proposed mechanism involves glycine uptake through glycine transporters 1 and 2 (GlyT 1–2) to reduce its concentration in the vicinity of β cells (Yan-Do et al. 2016). Glycine is then stored in insulin granules and synaptic-like microvesicles (Gammelsaeter et al. 2004), and released during exocytosis in a Ca^{2+}-dependent manner to act as an autocrine insulin secretion amplifier through glycine receptors (Gammelsaeter et al. 2004). Indeed, Strychnine, a potent and specific inhibitor of glycine receptors, abolishes glycine-induced GSIS amplification and transient inward currents in cells overexpressing GlyRα1 (Yan-Do et al. 2016). Beyond the autocrine effect of glycine, a single oral dose of glycine enhances insulin secretion when co-ingested with glucose in humans (González-Ortiz et al. 2001), which argues in favor of a role for glycine in regulating insulin secretion.

In humans, β-cell GlyRα1 and GlyT1,2 protein levels and circulating glycine levels are reduced in T2D (Floegel et al. 2013), suggesting a potential role of glycine and its downstream pathways in β-cell failure. On the other hand, glycine possesses antioxidant properties since it is a precursor of GSH, which is a reactive oxygen species scavenger (Sekhar et al. 2011). Despite this capacity, chronic glycine supplementation does not improve glycemic control or glucose tolerance in healthy or streptozotocin-induced T1D rat models (Chen et al. 2018). However, glycine supplementation increased circulating insulin during an oral glucose tolerance test in a T1D rat model (Chen et al. 2018), likely because of an observed β-cell mass increase, which opens up a new area of investigation into the role of glycine in improving β-cell function and survival.

3.2 *Regulation of Insulin Secretion by the Acetylcholine*

Intrapancreatic innervation by the parasympathetic efferent vagal nerve controls insulin secretion in response to feeding (Rodriguez-Diaz et al. 2011a). A vagotomy decreases GSIS (Siqueira et al. 2021), and efferent vagal stimulation potentiates insulin secretion (Frohman et al. 1967). Interestingly, GSIS decreased from islets isolated from rats 10 days after ex vivo vagotomy (Miranda et al. 2016), suggesting that islet parasympathetic innervation is essential to maintain β-cell function. Acetylcholine is the primary neurotransmitter of the parasympathetic system. Pancreatic α cells also produce and secrete acetylcholine, which acts as a paracrine regulator of insulin secretion (Rodriguez-Diaz et al. 2011b). M3 muscarinic receptor (M3R) is the most abundant isoform expressed in β cells. β-cell-specific M3R knockout mice are glucose intolerant and display impaired GSIS, in vivo and ex vivo (Gautam et al. 2006). Conversely, transgenic mice overexpressing M3R in β cells exhibit enhanced glucose tolerance and increased insulin secretion in both fed and fasted states (Gautam et al. 2006). The activation of M3R by endogenous ligands (i.e., acetylcholine) or by the muscarinic receptor agonist carbachol potentiates GSIS in human and rodent islets (Zawalich and Zawalich 2002; Sumara et al. 2009).

M3R, like the other $G_{\alpha q/11}$-coupled receptors detailed above (i.e., GPR40 and P2Y1), mediates its potentiating impacts through PKD1 (Sumara et al. 2009). However, the effects of PKD1 in M3R-dependent GSIS potentiation seem to bypass PLC and instead involve non-canonical β-arrestin-dependent PKD1 phosphorylation and activation (Kong et al. 2010) (Fig. 3). M3R activation also contributes to a rise in intracellular calcium levels that potentiates insulin secretion (Healy et al. 2010).

It is worth mentioning that because the $G_{\alpha q/11}$-coupled receptors described in this chapter potentiate GSIS with different amplitudes, multiple $G_{\alpha q/11}$-coupled receptor agonists are under investigation as new T2D treatments (Sebastiani et al. 2018). The observation that chronic activation of G_q-coupled receptors exclusively in β cells by a designer drug improves insulin secretion and protects against diabetes strengthens this hope (Jain et al. 2013).

3.3 *Insulin Secretion Regulation by the Incretin Hormones GLP-1 and GIP*

The amount of insulin secreted by β cells during a glucose tolerance test depends on the route of glucose administration. Oral glucose administration stimulates a higher insulin secretion response than intraperitoneal or intravenous injections (Elrick et al. 1964); this result is known as the incretin effect and is mediated by intestinal hormones secreted during intestinal nutrient transit. Glucagon-like peptide-1 (GLP-1) and glucose-dependent insulinotropic polypeptide (GIP) are the two main intestinal hormones mediating the incretin effect and the potentiation of GSIS.

GLP-1 is a hormone expressed and secreted by the enteroendocrine L-cells after sensing intestinal nutrient transition in the intestine (Reimann and Gribble 2002). GLP-1 is a cleavage product of proglucagon, which also produces glucagon and GLP2 (Bell et al. 1983). Although previous literature suggested that GLP-1 expression was limited to the intestine, increasing evidence suggests that GLP-1 is also expressed and secreted by pancreatic α cells (Campbell et al. 2021). Thus, GLP-1 acts as an endocrine and probably a paracrine hormone to potentiate GSIS (Campbell et al. 2021; Capozzi et al. 2019). β cells express high levels of the G-coupled GLP-1 receptor (GLP-1R) (Thorens 1992), which mediates the potentiation of insulin secretion by activating $G_{\alpha s}$ and $G_{\alpha q/11}$ downstream signaling pathways. $G_{\alpha s}$ activates adenylyl cyclase (AC) to synthesize the second messenger cAMP from ATP. Protein Kinase A (PKA) and the exchange protein activated by cAMP (EPAC) are activated downstream of cAMP and mediate the acute effects of GLP-1 on GSIS potentiation by regulating ion channel activities (Fig. 3). First, PKA (Light et al. 2002) and EPAC (Kang et al. 2006) close the K_{ATP} channel to promote plasma membrane depolarization. Next, the prolonged opening of cytoplasmic and reticulum endoplasmic calcium channels elevates intracellular calcium (Lu et al. 1993; Jacobo et al. 2009; Holz et al. 1999). Additional potentiation by GLP-1R stimulation involves the activation of exocytotic proteins (Ferdaoussi et al. 2020; Yasuda et al. 2010; Alenkvist et al. 2017).

GIP is another incretin hormone secreted by the proximal enteroendocrine K-cells (Parker et al. 2009) and exerts positive insulinotropic effects on β cells by activating GIP-R, a $G_{\alpha s}$-coupled receptor. GIP-dependent potentiation of GSIS involves the activation of AC, cAMP, and their downstream pathways, as discussed for GLP-1R activation (Seino et al. 2010) (Fig. 3). Interestingly, SENP1 is required to potentiate GSIS and exocytosis by GIP and GLP-1, probably by regulating exocytotic protein activity (Ferdaoussi et al. 2020; Lin et al. 2021).

A focus on GLP-1 and GIP studies diversifies therapeutic strategies to treat T2D, at least in part, by enhancing β-cell function, survival (Ferdaoussi et al. 2008), and regeneration (Ackeifi et al. 2020). Recently, the dual GLP-1R and GIP-R agonist Tirzepatide (Rosenstock et al. 2021), showed significant promise in treating T2D, an exciting step forward for this pernicious disease.

4 Conclusion Remarks

Multidisciplinary research agrees on the role of glucose and fatty acid metabolism on the amplification of insulin secretion; however, there is no ultimate consensus on which metabolic coupling factor mediates this amplification. Since these metabolic pathways act in cycles, favoring one metabolic factor in various experimental settings will shift the metabolic cycles toward different destinies. This possibility could explain the divergent findings. However, the molecular events involving SUMO/DeSUMOylation and the activation of exocytotic machinery appear to be the downstream targets of these metabolic coupling factors. Using advanced technology in vivo or investigating these mechanisms ex vivo by mimicking the in vivo extracellular milieu will help advance our understanding of the metabolic amplification of insulin secretion and reconcile the disparate interpretations (Alshamleh et al. 2020; Jensen et al. 2017). The G-protein-coupled receptors are another set of molecular mechanisms β cells use to potentiate insulin secretion. The advances in studying these receptors have greatly impacted the development of new therapeutic strategies to treat T2D. However, the β cell has not yet yielded its secrets on the molecular and metabolic mechanisms regulating insulin secretion amplification; understanding these mechanisms will guide further investigations to develop new T2D therapies preventing β-cell failure.

Acknowledgments For the memory of my dad, Ghali Ferdaoussi. Figures were generated in part using Servier Medical Art, provided by Servier, licensed under a Creative Commons Attribution 3.0 unported license.

Compliance with Ethical Standards Mourad Ferdaoussi declares no conflict of interest. This chapter does not contain any studies with any human participants or animals performed by the author.

References

Abulizi A, Cardone RL, Stark R, Lewandowski SL, Zhao X, Hillion J, Ma L, Sehgal R, Alves TC, Thomas C et al (2020) Multi-tissue acceleration of the mitochondrial phosphoenolpyruvate cycle improves whole-body metabolic health. Cell Metab 32:751–766.e11
Ackeifi C, Wang P, Karakose E, Manning Fox JE, González BJ, Liu H, Wilson J, Swartz E, Berrouet C, Li Y et al (2020) GLP-1 receptor agonists synergize with DYRK1A inhibitors to potentiate functional human β cell regeneration. Sci Transl Med 12:9996
Alenkvist I, Gandasi NR, Barg S, Tengholm A (2017) Recruitment of Epac2A to insulin granule docking sites regulates priming for exocytosis. Diabetes 66:2610–2622
Alquier T, Poitout V (2009) GPR40: good cop, bad cop? Diabetes 58:1035–1036
Alquier T, Peyot ML, Latour MG, Kebede M, Sorensen CM, Gesta S, Kahn CR, Smith RD, Jetton TL, Metz TO et al (2009) Deletion of GPR40 impairs glucose-induced insulin secretion in vivo in mice without affecting intracellular fuel metabolism in islets. Diabetes 58:2607–2615
Alshamleh I, Krause N, Richter C, Kurrle N, Serve H, Günther UL, Schwalbe H (2020) Real-time NMR spectroscopy for studying metabolism. Angew Chem Int Ed 59:2304–2308
Ang Z, Xiong D, Wu M, Ding JL (2018) FFAR2-FFAR3 receptor heteromerization modulates short-chain fatty acid sensing. FASEB J 32:289–303
Azzouny ME, Longacre MJ, Ansari IuH, Kennedy RT, Burant CF, MacDonald MJ (2016) Knockdown of ATP citrate lyase in pancreatic beta cells does not inhibit insulin secretion or glucose flux and implicates the acetoacetate pathway in insulin secretion. Mol Metab 5:980
Bauchle CJ, Rohli KE, Boyer CK, Pal V, Rocheleau JV, Liu S, Imai Y, Taylor EB, Stephens SB (2021) Mitochondrial efflux of citrate and isocitrate is fully dispensable for glucose-stimulated insulin secretion and pancreatic islet β-cell function. Diabetes 70:1717–1728
Bell GI, Santerre RF, Mullenbach GT (1983) Hamster preproglucagon contains the sequence of glucagon and two related peptides. Nature 302(5910):716–718
Bergeron V, Ghislain J, Poitout V (2016) The P21-activated kinase PAK4 is implicated in fatty-acid potentiation of insulin secretion downstream of free fatty acid receptor 1. Islets 8:157–164. https://doi.org/10.1080/19382014.2016.1243191
Bergeron V, Ghislain J, Vivot K, Tamarina N, Philipson LH, Fielitz J, Poitout V (2018) Deletion of protein kinase D1 in pancreatic β-cells impairs insulin secretion in high-fat diet-fed mice. Diabetes 67:71–77
Bhat EA, Sajjad N (2021) Human pannexin 1 channel: Insight in structure–function mechanism and its potential physiological roles. Mol Cell Biochem 476(3):1529–1540
Braun M, Wendt A, Karanauskaite J, Galvanovskis J, Clark A, MacDonald PE, Rorsman P (2007) Corelease and differential exit via the fusion pore of GABA, serotonin, and ATP from LDCV in rat pancreatic β cells. J Gen Physiol 129:221–231
Briscoe CP, Tadayyon M, Andrews JL, Benson WG, Chambers JK, Eilert MM, Ellis C, Elshourbagy NA, Goetz AS, Minnick DT et al (2003) The orphan G protein-coupled receptor GPR40 is activated by medium and long chain fatty acids. J Biol Chem 278:11303–11311
Brown JM, Betters JL, Lord C, Ma Y, Han X, Yang K, Alger HM, Melchior J, Sawyer J, Shah R et al (2010) CGI-58 knockdown in mice causes hepatic steatosis but prevents diet-induced obesity and glucose intolerance. J Lipid Res 51:3306–3315
Burnstock G (2018) Purine and purinergic receptors. Brain Neurosci Adv 2:2398212818817494
Campbell JE, Newgard CB (2021) Mechanisms controlling pancreatic islet cell function in insulin secretion. Nat Rev Mol Cell Biol 22:142–158
Campbell SA, Johnson J, Light PE (2021) Evidence for the existence and potential roles of intra-islet glucagon-like peptide-1. Islets 13:32–50. https://doi.org/10.1080/19382014.2021.1889941
Cantley J, Davenport A, Vetterli L, Nemes NJ, Whitworth PT, Boslem E, Thai LM, Mellett N, Meikle PJ, Hoehn KL et al (2019) Disruption of beta cell acetyl-CoA carboxylase-1 in mice impairs insulin secretion and beta cell mass. Diabetologia 62:99–111

Capozzi ME, Svendsen B, Encisco SE, Lewandowski SL, Martin MD, Lin H, Jaffe JL, Coch RW, Haldeman JM, MacDonald PE et al (2019) β Cell tone is defined by proglucagon peptides through cAMP signaling. JCI Insight 4(5):e126742
Chen L, Zhang J, Li C, Wang Z, Li J, Zhao D, Wang S, Zhang H, Huang Y, Guo X (2018) Glycine Transporter-1 and glycine receptor mediate the antioxidant effect of glycine in diabetic rat islets and INS-1 cells. Free Radic Biol Med 123:53–61
Craig TJ, Anderson D, Evans AJ, Girach F, Henley JM, Sudhof TC, Rizo J, Sudhof TC, Sollner T, Risselada HJ et al (2015) SUMOylation of Syntaxin1A regulates presynaptic endocytosis. Sci Rep 5:17669
Croze ML, Flisher MF, Guillaume A, Tremblay C, Noguchi GM, Granziera S, Vivot K, Castillo VC, Campbell SA, Ghislain J et al (2021) Free fatty acid receptor 4 inhibitory signaling in delta cells regulates islet hormone secretion in mice. Mol Metab 45:101166
Dai X-Q, Kolic J, Marchi P, Sipione S, Macdonald PE (2009) SUMOylation regulates Kv2.1 and modulates pancreatic beta-cell excitability. J Cell Sci 122:775–779
Dai X-Q, Plummer G, Casimir M, Kang Y, Hajmrle C, Gaisano HY, Manning Fox JE, MacDonald PE (2011) SUMOylation regulates insulin exocytosis downstream of secretory granule docking in rodents and humans. Diabetes 60:838–847
DiMeglio LA, Evans-Molina C, Oram RA (2018) Type 1 diabetes. Lancet 391:2449–2462
Dobbins RL, Chester MW, Stevenson BE, Daniels MB, Stein DT, McGarry JD (1998) A fatty acid-dependent step is critically important for both glucose- and non-glucose-stimulated insulin secretion. J Clin Invest 101:2370–2376
Eichmann TO, Lass A (2015) DAG tales: the multiple faces of diacylglycerol—stereochemistry, metabolism, and signaling. Cell Mol Life Sci 72:3931–3952
Elrick H, Stimmler L, Hlad CJ, Arai Y (1964) Plasma insulin response to oral and intravenous glucose administration. J Clin Endocrinol Metab 24:1076–1082
Ferdaoussi M, MacDonald PE (2017) Toward connecting metabolism to the exocytotic site. Trends Cell Biol 27(3):163–171
Ferdaoussi M, Abdelli S, Yang JY, Cornu M, Niederhauser G, Favre D, Widmann C, Regazzi R, Thorens B, Waeber G et al (2008) Exendin-4 protects β-cells from interleukin-1β-induced apoptosis by interfering with the c-Jun NH2-terminal kinase pathway. Diabetes 57:1205–1215
Ferdaoussi M, Bergeron V, Zarrouki B, Kolic J, Cantley J, Fielitz J, Olson EN, Prentki M, Biden T, MacDonald PE et al (2012) G protein-coupled receptor (GPR)40-dependent potentiation of insulin secretion in mouse islets is mediated by protein kinase D1. Diabetologia 55:2682–2692
Ferdaoussi M, Dai X, Jensen MV, Wang R, Peterson BS, Huang C, Ilkayeva O, Smith N, Miller N, Hajmrle C et al (2015) Isocitrate-to-SENP1 signaling amplifies insulin secretion and rescues dysfunctional β cells. J Clin Invest 125:3847–3860
Ferdaoussi M, Fu J, Dai X, Manning Fox JE, Suzuki K, Smith N, Plummer G, MacDonald PE (2017) SUMOylation and calcium control syntaxin-1A and secretagogin sequestration by tomosyn to regulate insulin exocytosis in human ß cells. Sci Rep 7(1):248
Ferdaoussi M, Smith N, Lin H, Bautista A, Spigelman AF, Lyon J, Dai XQ, Manning Fox JE, MacDonald PE (2020) Improved glucose tolerance with DPPIV inhibition requires β-cell SENP1 amplification of glucose-stimulated insulin secretion. Phys Rep 8:e14420
Floegel A, Stefan N, Yu Z, Mühlenbruch K, Drogan D, Joost HG, Fritsche A, Häring HU, De Angelis MH, Peters A et al (2013) Identification of serum metabolites associated with risk of type 2 diabetes using a targeted metabolomic approach. Diabetes 62:639–648
Frohman LA, Ezdinli EZ, Javid R (1967) Effect of vagotomy and vagal stimulation on insulin secretion. Diabetes 16:443–448
Gammelsaeter R, Frøyland M, Aragón C, Danbolt NC, Fortin D, Storm-Mathisen J, Davanger S, Gundersen V (2004) Glycine, GABA and their transporters in pancreatic islets of Langerhans: evidence for a paracrine transmitter interplay. J Cell Sci 117:3749–3758
Gautam D, Han SJ, Hamdan FF, Jeon J, Li B, Li JH, Cui Y, Mears D, Lu H, Deng C et al (2006) A critical role for β cell M3 muscarinic acetylcholine receptors in regulating insulin release and blood glucose homeostasis in vivo. Cell Metab 3:449–461

Gembal M, Detimary P, Gilon P, Gao ZY, Henquin JC (1993) Mechanisms by which glucose can control insulin release independently from its action on adenosine triphosphate-sensitive K+ channels in mouse B cells. J Clin Invest 91:871–880

Ghislain J, Poitout V (2021) Targeting lipid GPCRs to treat type 2 diabetes mellitus—progress and challenges. Nat Rev Endocrinol 17(3):162–175

Girach F, Craig TJ, Rocca DL, Henley JM (2013) RIM1α SUMOylation is required for fast synaptic vesicle exocytosis. Cell Rep 5:1294–1301

Goehring I, Sauter NS, Catchpole G, Assmann A, Shu L, Zien KS, Moehlig M, Pfeiffer AFH, Oberholzer J, Willmitzer L et al (2011) Identification of an intracellular metabolic signature impairing beta cell function in the rat beta cell line INS-1E and human islets. Diabetologia 54:2584–2594

González-Ortiz M, Medina-Santillán R, Martínez-Abundis E, Reynoso von Drateln C (2001) Effect of glycine on insulin secretion and action in healthy first-degree relatives of type 2 diabetes mellitus patients. Horm Metab Res 33:358–360

Gooding JR, Jensen MV, Dai X, Wenner BR, Lu D, Arumugam R, Ferdaoussi M, MacDonald PE, Newgard CB (2015) Adenylosuccinate is an insulin secretagogue derived from glucose-induced purine metabolism. Cell Rep 13:157–167

Guay C, Madiraju SRM, Aumais A, Joly É, Prentki M (2007) A role for ATP-citrate lyase, malic enzyme, and pyruvate/citrate cycling in glucose-induced insulin secretion. J Biol Chem 282:35657–35665

Guay C, Joly É, Pepin É, Barbeau A, Hentsch L, Pineda M, Madiraju SRM, Brunengraber H, Prentki M (2013) A role for cytosolic isocitrate dehydrogenase as a negative regulator of glucose signaling for insulin secretion in pancreatic ß-cells. PLoS One 8:e77097

Hauke S, Keutler K, Phapale P, Yushchenko DA, Schultz C (2018) Endogenous fatty acids are essential signaling factors of pancreatic β-cells and insulin secretion. Diabetes 67:1986–1998

Healy JA, Nilsson KR, Hohmeier HE, Berglund J, Davis J, Hoffman J, Kohler M, Li LS, Berggren PO, Newgard CB et al (2010) Cholinergic augmentation of insulin release requires ankyrin-B. Sci Signal 3(113):ra19

Henquin JC (2000) Triggering and amplifying pathways of regulation of insulin secretion by glucose. Diabetes 49:1751–1760

Holz GG, Leech CA, Scott Heller R, Castonguay M, Habener JF (1999) cAMP-dependent mobilization of intracellular Ca2+ stores by activation of ryanodine receptors in pancreatic β-cells: a Ca2+ signaling system stimulated by the insulinotropic hormone glucagon-like peptide-1-(7–37). J Biol Chem 274:14147–14156

Ishihara H (2022) Metabolism-secretion coupling in glucose-stimulated insulin secretion. Diabetol Int 13:463–470

Ivarsson R, Quintens R, Dejonghe S, Tsukamoto K, In't Veld P, Renström E, Schuit FC (2005) Redox control of exocytosis: regulatory role of NADPH, thioredoxin, and glutaredoxin. Diabetes 54:2132–2142

Jacobo SMP, Guerra ML, Hockerman GH (2009) Cav1.2 and Cav1.3 are differentially coupled to glucagon-like peptide-1 potentiation of glucose-stimulated insulin secretion in the pancreatic β-cell line INS-1. J Pharmacol Exp Ther 331:724–732

Jacques-Silva MC, Correa-Medina M, Cabrera O, Rodriguez-Diaz R, Makeeva N, Fachado A, Diez J, Berman DM, Kenyon NS, Ricordi C et al (2010) ATP-gated P2X3 receptors constitute a positive autocrine signal for insulin release in the human pancreatic β cell. Proc Natl Acad Sci USA 107:6465

Jain S, De AIR, Lu H, White MF, Guettier JM, Wess J (2013) Chronic activation of a designer Gq-coupled receptor improves β cell function. J Clin Invest 123:1750–1762

Jansen NS, Vertegaal ACO (2021) A chain of events: regulating target proteins by SUMO polymers. Trends Biochem Sci 46:113–123

Jensen MV, Gooding JR, Ferdaoussi M, Dai X-Q, Peterson BS, MacDonald PE, Newgard CB (2017) Metabolomics applied to islet nutrient sensing mechanisms. Diabetes Obes Metab 19(Suppl 1):90–94

Joseph JW, Jensen MV, Ilkayeva O, Palmieri F, Alárcon C, Rhodes CJ, Newgard CB (2006) The mitochondrial citrate/isocitrate carrier plays a regulatory role in glucose-stimulated insulin secretion. J Biol Chem 281:35624–35632

Just T, Pau HW, Engel U, Hummel T (2008) Cephalic phase insulin release in healthy humans after taste stimulation? Appetite 51:622–627

Kaku K, Enya K, Nakaya R, Ohira T, Matsuno R (2015) Efficacy and safety of fasiglifam (TAK-875), a G protein-coupled receptor 40 agonist, in Japanese patients with type 2 diabetes inadequately controlled by diet and exercise: a randomized, double-blind, placebo-controlled, phase III trial. Diabetes Obes Metab 17:675–681

Kang G, Chepurny OG, Malester B, Rindler MJ, Rehmann H, Bos JL, Schwede F, Coetzee WA, Holz GG (2006) cAMP sensor Epac as a determinant of ATP-sensitive potassium channel activity in human pancreatic β cells and rat INS-1 cells. J Physiol 573:595–609

Kebede M, Alquier T, Latour MG, Semache M, Tremblay C, Poitout V (2008) The fatty acid receptor GPR40 plays a role in insulin secretion in vivo after high-fat feeding. Diabetes 57:2432–2437

Khan S, Yan-Do R, Duong E, Wu X, Bautista A, Cheley S, Macdonald PE, Braun M (2014) Autocrine activation of P2Y1 receptors couples Ca2+ influx to Ca2+ release in human pancreatic beta cells. Diabetologia 57:2535–2545

Khan S, Ferdaoussi M, Bautista A, Bergeron V, Smith N, Poitout V, MacDonald PE (2019) A role for PKD1 in insulin secretion downstream of $P2Y_1$ receptor activation in mouse and human islets. Phys Rep 7(19):e14250

Kibbey RG, Pongratz RL, Romanelli AJ, Wollheim CB, Cline GW, Shulman GI (2007) Mitochondrial GTP regulates glucose-stimulated insulin secretion. Cell Metab 5:253–264

Kong KC, Butcher AJ, McWilliams P, Jones D, Wess J, Hamdan FF, Werry T, Rosethorne EM, Charlton SJ, Munson SE et al (2010) M3-muscarinic receptor promotes insulin release via receptor phosphorylation/arrestin-dependent activation of protein kinase D1. Proc Natl Acad Sci USA 107:21181–21186

Kyriazis GA, Soundarapandian MM, Tyrberg B (2012) Sweet taste receptor signaling in beta cells mediates fructose-induced potentiation of glucose-stimulated insulin secretion. Proc Natl Acad Sci USA 109:E524

Latour MG, Alquier T, Oseid E, Tremblay C, Jetton TL, Luo J, Lin DC-H, Poitout V (2007) GPR40 is necessary but not sufficient for fatty acid stimulation of insulin secretion in vivo. Diabetes 56:1087–1094

Léon C, Freund M, Latchoumanin O, Farret A, Petit P, Cazenave JP, Gachet C (2005) The P2Y1 receptor is involved in the maintenance of glucose homeostasis and in insulin secretion in mice. Purinergic Signal 1:145–151

Light PE, Manning Fox JE, Riedel MJ, Wheeler MB (2002) Glucagon-like peptide-1 inhibits pancreatic ATP-sensitive potassium channels via a protein kinase A- and ADP-dependent mechanism. Mol Endocrinol 16:2135–2144

Lin H, Smith N, Spigelman AF, Suzuki K, Ferdaoussi M, Alghamdi TA, Lewandowski SL, Jin Y, Bautista A, Wang YW et al (2021) β-cell knockout of SENP1 reduces responses to incretins and worsens oral glucose tolerance in high-fat diet–fed mice. Diabetes 70:2626–2638

Lu M, Wheeler MB, Leng X-H, Boyd AE (1993) The role of the free cytosolic calcium level in beta-cell signal transduction by gastric inhibitory polypeptide and glucagon-like peptide I(7-37). Endocrinology 132:94–100

Lu D, Mulder H, Zhao P, Burgess SC, Jensen MV, Kamzolova S, Newgard CB, Sherry AD (2002) 13C NMR isotopomer analysis reveals a connection between pyruvate cycling and glucose-stimulated insulin secretion (GSIS). Proc Natl Acad Sci USA 99:2708–2713

Macdonald MJ, Brown LJ, Longacre MJ, Stoker SW, Kendrick MA (2013) Knockdown of both mitochondrial isocitrate dehydrogenase enzymes in pancreatic beta cells inhibits insulin secretion. Biochim Biophys Acta Gen Subj 1830:5104–5111

Mancini AD, Bertrand G, Vivot K, Carpentier É, Tremblay C, Ghislain J, Bouvier M, Poitout V (2015) β-arrestin recruitment and biased agonism at free fatty acid receptor 1. J Biol Chem 290:21131–21140
Martins EF, Miyasaka CK, Newsholme P, Curi R, Carpinelli AR (2004) Changes of fatty acid composition in incubated rat pancreatic islets: evidence for fatty acid release. Diabetes Metab 30:21–27
McNelis JC, Lee YS, Mayoral R, Van Der Kant R, Johnson AMF, Wollam J, Olefsky JM (2015) GPR43 potentiates β-cell function in obesity. Diabetes 64:3203–3217
Merrins MJ, Corkey BE, Kibbey RG, Prentki M (2022) Metabolic cycles and signals for insulin secretion. Cell Metab 34(7):947–968
Miranda RA, Torrezan R, de Oliveira JC, Barella LF, da Silva Franco CC, Lisboa PC, Moura EG, Mathias PCF (2016) HPA axis and vagus nervous function are involved in impaired insulin secretion of MSG-obese rats. J Endocrinol 230:27–38
Mugabo Y, Zhao S, Seifried A, Gezzar S, Al-Mass A, Zhang D, Lamontagne J, Attane C, Poursharifi P, Iglesias J et al (2016) Identification of a mammalian glycerol-3-phosphate phosphatase: Role in metabolism and signaling in pancreatic β-cells and hepatocytes. Proc Natl Acad Sci USA 113:E430–E439
Nishizaki H, Matsuoka O, Kagawa T, Kobayashi A, Watanabe M, Moritoh Y (2021) SCO-267, a GPR40 full agonist, stimulates islet and gut hormone secretion and improves glycemic control in humans. Diabetes 70:2364–2376
Parker HE, Habib AM, Rogers GJ, Gribble FM, Reimann F (2009) Nutrient-dependent secretion of glucose-dependent insulinotropic polypeptide from primary murine K cells. Diabetologia 52:289–298
Pongratz RL, Kibbey RG, Shulman GI, Cline GW (2007) Cytosolic and mitochondrial malic enzyme isoforms differentially control insulin secretion. J Biol Chem 282:200–207
Prentki M, Nolan CJ (2006) Islet beta cell failure in type 2 diabetes. J Clin Invest 116:1802–1812
Prentki M, Vischer S, Glennon MC, Regazzi R, Deeney JT, Corkey BE (1992) Malonyl-CoA and long chain acyl-CoA esters as metabolic coupling factors in nutrient-induced insulin secretion. J Biol Chem 267:5802–5810
Prentki M, Corkey BE, Madiraju SRM (2020) Lipid-associated metabolic signalling networks in pancreatic beta cell function. Diabetologia 63:10–20
Priyadarshini M, Villa SR, Fuller M, Wicksteed B, Mackay CR, Alquier T, Poitout V, Mancebo H, Mirmira RG, Gilchrist A et al (2015) An acetate-specific GPCR, FFAR2, regulates insulin secretion. Mol Endocrinol 29:1055–1066
Reimann F, Gribble FM (2002) Glucose-sensing in glucagon-like peptide-1-secreting cells. Diabetes 51:2757–2763
Rhee JS, Betz A, Pyott S, Reim K, Varoqueaux F, Augustin I, Hesse D, Südhof TC, Takahashi M, Rosenmund C et al (2002) β phorbol ester- and diacylglycerol-induced augmentation of transmitter release is mediated by Munc13s and not by PKCs. Cell 108:121–133
Rodriguez-Diaz R, Abdulreda MH, Formoso AL, Gans I, Ricordi C, Berggren PO, Caicedo A (2011a) Innervation patterns of autonomic axons in the human endocrine pancreas. Cell Metab 14:45–54
Rodriguez-Diaz R, Dando R, Jacques-Silva MC, Fachado A, Molina J, Abdulreda MH, Ricordi C, Roper SD, Berggren PO, Caicedo A (2011b) Alpha cells secrete acetylcholine as a non-neuronal paracrine signal priming beta cell function in humans. Nat Med 17(7):888–892
Roduit R, Nolan C, Alarcon C, Moore P, Barbeau A, Delghingaro-Augusto V, Przybykowski E, Morin J, Massé F, Massie B et al (2004) A role for the malonyl CoA/long-chain acyl-CoA pathway of lipid signaling in the regulation of insulin secretion in response to both fuel and nonfuel stimuli. Diabetes 53:1007–1019
Ronnebaum SM, Ilkayeva O, Burgess SC, Joseph JW, Lu D, Stevens RD, Becker TC, Sherry AD, Newgard CB, Jensen MV (2006) A pyruvate cycling pathway involving cytosolic NADP-dependent isocitrate dehydrogenase regulates glucose-stimulated insulin secretion. J Biol Chem 281:30593–30602

Ronnebaum SM, Jensen MV, Hohmeier HE, Burgess SC, Zhou YP, Qian S, MacNeil D, Howard A, Thornberry N, Ilkayeva O et al (2008) Silencing of cytosolic or mitochondrial isoforms of malic enzyme has no effect on glucose-stimulated insulin secretion from rodent islets. J Biol Chem 283:28909–28917

Rosenstock J, Wysham C, Frías JP, Kaneko S, Lee CJ, Fernández Landó L, Mao H, Cui X, Karanikas CA, Thieu VT (2021) Efficacy and safety of a novel dual GIP and GLP-1 receptor agonist tirzepatide in patients with type 2 diabetes (SURPASS-1): a double-blind, randomised, phase 3 trial. Lancet 398:143–155

Sabrautzki S, Kaiser G, Przemeck GKH, Gerst F, Lorza-Gil E, Panse M, Sartorius T, Hoene M, Marschall S, Häring HU et al (2017) Point mutation of Ffar1 abrogates fatty acid-dependent insulin secretion, but protects against HFD-induced glucose intolerance. Mol Metab 6:1304

Sato Y, Aizawa T, Komatsu M, Okada N, Yamada T (1992) Dual functional role of membrane depolarization/Ca2+ influx in rat pancreatic B-cell. Diabetes 41:438–443

Schuit F, De VA, Farfari S, Moens K, Pipeleers D, Brun T, Prentki M (1997) Metabolic fate of glucose in purified islet cells: glucose-regulated anaplerosis in β cells. J Biol Chem 272:18572–18579

Sebastiani G, Ceccarelli E, Castagna MG, Dotta F (2018) G-protein-coupled receptors (GPCRs) in the treatment of diabetes: Current view and future perspectives. Best Pract Res Clin Endocrinol Metab 32:201–213

Seino Y, Fukushima M, Yabe D (2010) GIP and GLP-1, the two incretin hormones: Similarities and differences. J Diabetes Invest 1:8–23

Sekhar RV, Mckay SV, Patel SG, Guthikonda AP, Reddy VT, Balasubramanyam A, Jahoor F (2011) Glutathione synthesis is diminished in patients with uncontrolled diabetes and restored by dietary supplementation with cysteine and glycine. Diabetes Care 34:162–167

Serrano J, Meshram NN, Soundarapandian MM, Smith KR, Mason C, Brown IS, Tyrberg B, Kyriazis GA (2022) Saccharin stimulates insulin secretion dependent on sweet taste receptor-induced activation of PLC signaling axis. Biomedicines 10:120

Sims K (2021) Congenital Hyperinsulinism. NeoReviews 22:e230–e240

Siqueira BS, Ceglarek VM, Gomes ECZ, Vettorazzi JF, Rentz T, Nenevê JZ, Volinski KZ, Moraes SS, Malta A, De Freitas Mathias PC et al (2021) Vagotomy and splenectomy reduce insulin secretion and interleukin-1ß. Pancreas 50:607–616

Spégel P, Sharoyko VV, Goehring I, Danielsson APH, Malmgren S, Nagorny CLF, Andersson LE, Koeck T, Sharp GWG, Straub SG et al (2013) Time-resolved metabolomics analysis of β-cells implicates the pentose phosphate pathway in the control of insulin release. Biochem J 450:595–605

Sumara G, Formentini I, Collins S, Sumara I, Windak R, Bodenmiller B, Ramracheya R, Caille D, Jiang H, Platt KA et al (2009) Regulation of PKD by the MAPK p38δ in insulin secretion and glucose homeostasis. Cell 136:235–248

Syed SK, Kauffman AL, Beavers LS, Alston JT, Farb TB, Ficorilli J, Marcelo MC, Brenner MB, Bokvist K, Barrett DG et al (2013) Ectonucleotidase NTPDase3 is abundant in pancreatic β-cells and regulates glucose-induced insulin secretion. Am J Physiol Endocrinol Metab 305(10):E1319–E1326

Tang C, Ahmed K, Gille A, Lu S, Gröne HJ, Tunaru S, Offermanns S (2015) Loss of FFA2 and FFA3 increases insulin secretion and improves glucose tolerance in type 2 diabetes. Nat Med 21(2):173–177

Thorens B (1992) Expression cloning of the pancreatic beta cell receptor for the gluco-incretin hormone glucagon-like peptide 1. Proc Natl Acad Sci USA 89:8641–8645

Tozzi M, Larsen AT, Lange SC, Giannuzzo A, Andersen MN, Novak I (2018) The P2X7 receptor and pannexin-1 are involved in glucose-induced autocrine regulation in β-cells. Sci Rep 8(1):1–15

Trauelsen M, Lückmann M, Frimurer TM, Schwartz TW (2018) The HETE is on FFAR1 and pancreatic islet cells. Cell Metab 27:273–275

Usui R, Yabe D, Fauzi M, Goto H, Botagarova A, Tokumoto S, Tatsuoka H, Tahara Y, Kobayashi S, Manabe T et al (2019) GPR40 activation initiates store-operated Ca2+ entry and potentiates insulin secretion via the IP3R1/STIM1/Orai1 pathway in pancreatic β-cells. Sci Rep 9(1):1–11
Wuttke A, Idevall-Hagren O, Tengholm A (2013) P2Y1 receptor-dependent diacylglycerol signaling microdomains in β cells promote insulin secretion. FASEB J 27:1610–1620
Xu J, Han J, Long YS, Lock J, Weir GC, Epstein PN, Liu YQ (2008) Malic enzyme is present in mouse islets and modulates insulin secretion. Diabetologia 51:2281–2289
Yamada H, Yoshida M, Ito K, Dezaki K, Yada T, Ishikawa SE, Kakei M (2016) Potentiation of glucose-stimulated insulin secretion by the GPR40–PLC–TRPC pathway in pancreatic β-cells. Sci Rep 6(1):1–9
Yan-Do R, Duong E, Manning Fox JE, Dai X, Suzuki K, Khan S, Bautista A, Ferdaoussi M, Lyon J, Wu X et al (2016) A glycine-insulin autocrine feedback loop enhances insulin secretion from human β-cells and is impaired in type 2 diabetes. Diabetes 65(8):2311–2321
Yashiro H, Tsujihata Y, Takeuchi K, Hazama M, Johnson PRV, Rorsman P (2012) The effects of TAK-875, a selective G protein-coupled receptor 40/free fatty acid 1 agonist, on insulin and glucagon secretion in isolated rat and human islets. J Pharmacol Exp Ther 340:483–489
Yasuda T, Shibasaki T, Minami K, Takahashi H, Mizoguchi A, Uriu Y, Numata T, Mori Y, Miyazaki JI, Miki T et al (2010) Rim2α determines docking and priming states in insulin granule exocytosis. Cell Metab 12:117–129
Zawalich WS, Zawalich KC (2002) Effects of glucose, exogenous insulin, and carbachol on C-peptide and insulin secretion from isolated perifused rat islets. J Biol Chem 277:26233–26237
Zhang GF, Jensen MV, Gray SM, El K, Wang Y, Lu D, Becker TC, Campbell JE, Newgard CB (2021) Reductive TCA cycle metabolism fuels glutamine- and glucose-stimulated insulin secretion. Cell Metab 33:804–817.e5
Zhao S, Mugabo Y, Iglesias J, Xie L, Delghingaro-Augusto V, Lussier R, Peyot ML, Joly E, Taïb B, Davis MA et al (2014) α/β-hydrolase domain-6-accessible monoacylglycerol controls glucose-stimulated insulin secretion. Cell Metab 19:993–1007

Pancreatic δ Cells: An Overlooked Cell in Focus

Maria L. Golson

Abstract Pancreatic δ cells act locally to repress both insulin and glucagon secretion. Because they are a rare cell type, experimentation examining δ-cell function and control has lagged that of the more abundant α and β cells. Emerging evidence, enabled partly by developing single-cell technology, demonstrates that δ-cell function is, in part, directed by δ cells but that δ cells also have intrinsic control. The contribution of these cells to overall glucose homeostasis and diabetes onset and progression is still unclear. However, they regulate both α and β cells, both of which are dysfunctional in diabetes, and their numbers are disrupted in humans with diabetes and in multiple animal models of diabetes, suggesting δ cells are a pivotal character in both health and disease.

Keywords δ Cell · Islets · Somatostatin · Diabetes

Abbreviations

Abcc8	ATP-binding cassette subfamily C member 8
Adcy	Adenylate cyclase
cAMP	3′,5′ Cyclic adenosine monophosphate
Epac2	Rap guanine nuclear exchange factor 2
FACS	Fluorescence-activated cell sorting
GLP-1	Glucagon-like peptide 1
HHEX	Hematopoietically expressed homeobox
KCNA5	Potassium voltage-gated channel subfamily A member 5
KCNJ1	Potassium inwardly-rectifying channel subfamily J member 1

M. L. Golson (✉)
Division of Endocrinology, Diabetes, and Metabolism, Department of Medicine, Johns Hopkins University, Baltimore, MD, USA
e-mail: mgolson1@jhmi.edu

M. L. Golson (ed.), *Molecular and Cell Biology of Pancreas Development, Function and Regeneration*, Advances in Anatomy, Embryology and Cell Biology 239, https://doi.org/10.1007/978-3-031-62232-8_6

LEPR	Leptin receptor
NGN3	Neurogenin3
PKA	Protein kinase A
RNAseq	RNA sequencing
SLC	Solute carrier
SST	Somatostatin
SSTR	Somatostatin receptor
T1D	Type 1 diabetes
T2D	Type 2 diabetes

1 Introduction

The islets of Langerhans consist of several different cell types that work together to control glucose homeostasis, the large bulk of which are α, β, and δ cells (Cabrera et al. 2006). β Cells release insulin when glucose is high. Insulin acts by inducing glucose uptake and utilization in muscles, lipogenesis in adipose tissue, and decreased glucose output by the liver (reviewed in Tyagi and Pugazhenthi 2021). When glucose is low, α cells secrete glucagon, a hormone that instructs the liver to release glycogen stores and induces gluconeogenesis from amino acids (Finan et al. 2020). These two cell types also release other factors that can influence the activity of other islet cells. For example, α cells release acetylcholine, a signal that activates α, β, and δ cells (Rodriguez-Diaz et al. 2011; Duttaroy et al. 2004; Honey and Weir 1980). Similarly, β cells release Urocortin 3, which triggers somatostatin (SST) release from the δ cell (van der Meulen et al. 2015; Li et al. 2018).

δ cells primarily attenuate the activity of other islet endocrine cells. δ cells produce somatostatin, an autocrine and paracrine hormone that inhibits secretory cells (Rorsman and Huising 2018; Shamsi et al. 2021). In the islet, somatostatin represses α and β cells (Strowski et al. 2000; Hauge-Evans et al. 2009), and likely δ, pancreatic polypeptide, and ε cells (Shamsi et al. 2021). In the adult mouse, δ cells make up 2–5% of the islet and tend to lie at the islet periphery, along with other non-β cells (Steiner et al. 2010). δ cells make up between 5 and 10% of the adult human islet (Wang et al. 2016a) and are interspersed throughout the islet, which has raised speculation that intra-islet communication has a more crucial role in regulating β-cell function in human islets compared to the widely used mouse model.

This chapter will discuss (1) how somatostatin functions, (2) how δ cells contribute to blood glucose homeostasis, (3) how differences in structure between mouse and human islets might affect communications between islet cells, (4) how δ cells are activated, (5) whether molecules that regulate β-cell function other than somatostatin are released from δ cells, and (6) how δ cells might prevent or contribute to diabetes onset and progression.

2 Advances and Challenges in δ-Cell Experimentation

δ cells make up only a small portion of the islet and are intermingled with other islet cell types, regardless of the species examined (Steiner et al. 2010). For these reasons, the study of the δ cell has lagged that of α and β cells. Several newer technologies and new implementations of older technologies are allowing for the elucidation of the role δ cells play in glucose homeostasis and the mechanistic control of δ cell function. Single-cell assays like RNAseq and protein mass cytometry, the development of antibodies for sorting human islet cells, and the ability to express fluorescent reporters in mouse δ cells for fluorescence-activated cell sorting (FACS) have permitted the identification of genes and proteins expressed in δ cells (Wang et al. 2016a; DiGruccio et al. 2016; Dorrell et al. 2008; Kaestner et al. 2019). Moreover, genetic mouse models have allowed for the expression of cAMP reporters or the examination of membrane currents or Ca^{2+} entry specifically in δ cells (Denwood et al. 2019).

Despite these advances, challenges remain in understanding δ cells. First, there is currently no genetic system that allows gene expression or ablation specifically in pancreatic δ cells. Current approaches rely on the somatostatin promoter to drive gene expression or ablation (Cigliola et al. 2018; Taniguchi et al. 2011). Since somatostatin is expressed in brain and gut enteroendocrine cells, expression or deletion of genes using these models affects all these metabolic tissues, not just pancreatic δ cells (Ampofo et al. 2020). This shortcoming can lead to difficulty in interpreting experimental outcomes. For example, *Sst*$^{-/-}$ mice survive and display mostly normal glucose homeostasis, albeit with decreased blood glucose excursions after a glucose bolus and increased secretion from isolated islets (Hauge-Evans et al. 2009, 2015; Zeyda et al. 2001). Mice that express diphtheria toxin under the control of the somatostatin promoter (*Sst*^{DTA} mice), however, have a much more severe phenotype (Li et al. 2018). This system results in the ablation of all cells that express somatostatin. Mice with this genetic alteration die within 24 h of birth; this short survival time can be increased by approximately 48 h with glucose supplementation, supporting a contributory role for disturbed glucose homeostasis in the mortality of these mice.

The discrepancy between phenotypes when somatostatin-expressing cells are destroyed versus the milder presentation when only somatostatin is ablated may insinuate that the role of δ cells is far beyond that of just somatostatin release. However, it is impossible to attribute the lethality of mice lacking somatostatin-expressing cells solely to the pancreatic δ cells since somatostatin is expressed so widely. Indeed, mice lacking pancreatic endocrine *Hematopoietically Expressed Homeobox (Hhex)* (*Neurog3-Cre;Hhex*$^{loxP/loxP}$ mice), a master transcriptional regulator of δ-cell development and maintenance, fail to develop δ cells but survive to adulthood with increased insulin and glucagon secretion (Zhang et al. 2014). The survival of *Ngn3-Cre; Hhex*$^{loxP/loxP}$ mice suggests that the loss of somatostatin-expressing cells in an extrapancreatic location causes lethality in *Sst*^{DTA} mice.

Other significant impediments to examining δ cell function include: (1) a lack of a characterized human δ cell-specific promoter and (2) the inability of even the most sophisticated live human islet-sorting protocol to distinguish β from δ cells (Saunders et al. 2019).

3 Somatostatin Use in Medicine

Because of its broad role in inhibiting secretory cells, somatostatin modulation is of interest in medicine. Because somatostatin has a short half-life of only minutes (Benuck and Marks 1976), the hormone is not a suitable treatment candidate. Somatostatin analogs were first approved for treating certain diseases in the 1980s; they have a much longer half-life than exogenous somatostatin, ranging from 90 min to 4.5 days for slow-release versions (Gomes-Porras et al. 2020). Somatostatin analogs are currently used in the treatment of several diseases and disorders, including carcinoid, breast, colon, prostate, lung, exocrine pancreas, and hepatocellular tumors, pituitary adenomas, and acromegaly—a condition characterized by hypersecretion of growth hormone during adulthood– thyrotropinomas, glucagonomas, and hyperinsulinemic hypoglycemia (Gomes-Porras et al. 2020; Haris et al. 2020; Stueven et al. 2019). Many of these uses are off-label. In addition, somatostatin analogs are also used to treat diabetic retinopathy and diabetic macular edema since somatostatin is produced in the retina and decreased in the diabetic state (Hernandez et al. 2014). Treatment with somatostatin analogs, in some cases, can result in diabetes until drug withdrawal (Cappellani et al. 2019), highlighting the importance of δ-cell regulation in maintaining blood glucose homeostasis.

4 Somatostatin Action

4.1 Somatostatin and Its Receptors

Somatostatin is produced as a 116 amino acid preprohormone and processed into either a 28 amino acid (SST-28) or 14 amino acid (SST-14) bioactive form, the latter of which is the primary cleavage product in the pancreatic δ cell (Francis et al. 1990). Somatostatin represses hormone secretion through somatostatin receptor (SSTR) signaling (Fig. 1). Five somatostatin receptors have been identified. Receptors 1–4 have similar binding affinities for SST-14 and SST-28, but SSTR-5 has a 5–10-fold greater binding coefficient for SST-28 than SST-14. Somatostatin is produced not only by pancreatic δ cells but also by cells elsewhere in the body, including the hypothalamus, peripheral neurons, the adrenal gland, and gut enteroendocrine cells. Most circulating somatostatin is produced by gut enteroendocrine cells (Colturi et al. 1984; Loud et al. 1985; Huising et al. 2018). However, because

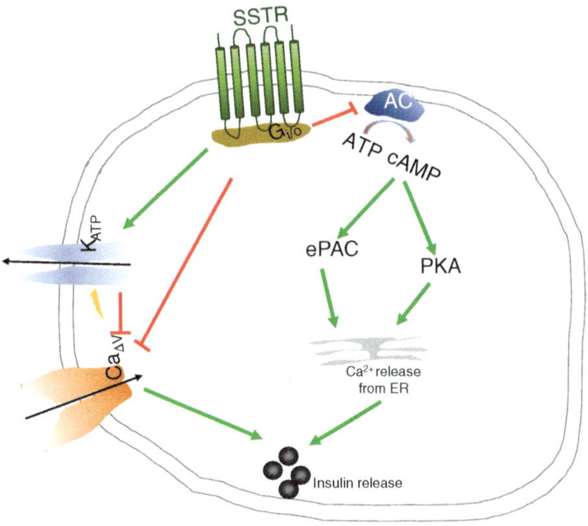

Fig. 1 SSTR signaling inhibits hormone secretion by multiple methods mediated by inhibitory G proteins. In β cells, G proteins activate the ATP-sensitive K^+ channel and close the voltage-gated Ca^{2+} channel. In addition, SSTR signaling inhibits Ca^{2+} release from the endoplasmic reticulum

somatostatin has a short half-life in the blood of only 1–3 min (Benuck and Marks 1976), circulating somatostatin is too dilute to activate islet somatostatin receptors (Huising et al. 2018). Taken together, these data strongly suggest that only or primarily local somatostatin depresses islet hormone release and that little, if any, somatostatin suppression of islet function is due to gut or neuronal sources.

4.2 *Somatostatin Receptor Expression Patterns*

Different islet cells have varying levels of signaling through the five distinct somatostatin receptors, and divergent use of the receptors within at least some cell types is apparent between species. However, α cell SSTR expression and utilization are consistent, at least among mice, rats, and humans. Experiments employing receptor-specific agonists and antagonists, knockouts, and antibody labeling have demonstrated that SSTR2 relays most of the somatostatin signal in mouse, rat, and human α cells (Hunyady et al. 1997; Kumar et al. 1999; Kailey et al. 2012).

In contrast to α cells, somatostatin receptor expression and utilization in β cells differ between mice and humans. Immunolabeling studies demonstrate that human β cells express SSTR1, 2, and 5. *SSTR3* RNA is also present (Kumar et al. 1999). Functional studies illustrate that, like human α cells, SSTR2 predominates as the mediator of somatostatin action in human β cells (Kailey et al. 2012; Singh et al. 2007). Mouse and human β cells display a strikingly distinct difference in SSTR expression. In mouse β cells, studies using SSTR agonists suggest that SSTR5 is the primary mediator of somatostatin signaling (Strowski et al. 2000, 2003; Rohrer and Schaeffer 2000). In addition, mice null for *Sstr5* exhibit elevated blood insulin levels; furthermore, islets from these mice display increased insulin secretion in response to high glucose and a blunted response to SST-28 (Strowski et al. 2000). Confusingly, however, no *SSTR5* RNA expression is detectable by RNAseq in mouse α, β, or δ cells, and, in mouse β cells, of all the SSTR receptors, only SSTR3 is detectable at any appreciable level (DiGruccio et al. 2016).

In human δ cells, *SSTR1*, *SSTR2*, and *SSTR3* have similar RNA expression levels, which are higher than those of *SSTR4* and *SSTR5* (Wang et al. 2016b). Immunolabeling, however, pinpoints SSTR5 as the SSTR expressed in the highest percentage of δ cells (Kumar et al. 1999). In mouse δ cells, *Sstr3* RNA is elevated compared to other somatostatin receptors (Adriaenssens et al. 2016); SSTR protein expression and functional data from mouse δ cells are lacking.

4.3 Downstream Effects of Somatostatin Receptor Activation in Islet Endocrine Cells

All SSTRs are associated with inhibitory G_i or G_o proteins, and somatostatin binding to SSTRs activates their associated G proteins (Bo et al. 2022). The most well-studied downstream action of SSTR signaling is adenylate cyclase (ADCY) inhibition, which results in reduced cAMP (Patel et al. 1994). In β cells, cAMP acts in the amplifying pathway to enhance glucose-stimulated insulin secretion (Kalwat and Cobb 2017). cAMP activates protein kinase A (PKA) and Rap guanine nuclear exchange factor 2 (Epac2), both of which evoke Ca^{2+} release from endoplasmic reticulum stores; therefore, ADCY inhibition attenuates the amplifying pathway of insulin secretion (Kaihara et al. 2013; Holz 2004). The absence of somatostatin increases insulin secretion during both the first and the second phases of GSIS, but not basal insulin secretion, as demonstrated by the perifusion of islets isolated from mice deficient in somatostatin (Hauge-Evans et al. 2009). The triggering pathway has historically been associated with the first phase of insulin secretion, and the amplifying pathway with the second. More recent evidence suggests that this distinction is inaccurate and that the amplifying pathway acts in both phases to enhance insulin secretion (Kalwat and Cobb 2017). This more recent interpretation is consistent with somatostatin working through SSTR to decrease cAMP, and thereby insulin secretion, during both phases of insulin secretion. In addition to changes in cAMP, SSTR downstream action also counteracts glucose-stimulated insulin secretion by activating inwardly rectifying potassium channels and suppressing voltage-gated calcium channels in β cells (Kailey et al. 2012).

The mechanisms that control glucagon secretion are less understood than those for insulin. Glucagon is typically described as a hormone secreted at low glucose levels, but glucagon secretion has been described as following a U-shaped curve, with peaks in secretion from mouse islets at both 0 and 25 mM glucose and a nadir around 7 mM glucose (Gylfe 2016). These data suggest that α cells may respond to multiple intrinsic and extrinsic signals with opposing actions that alternatively activate or suppress glucagon secretion. Further, it implies that somatostatin can repress glucagon secretion at medium glucose levels. In contrast, somatostatin cannot countermand signaling events after glucose exceeds a certain threshold and aberrantly activates α cells.

Both cAMP and Ca^{2+} are essential messengers that positively correlate with α-cell glucagon secretion (Tengholm and Gylfe 2017). In contrast to β cells, cAMP may be more crucial than Ca^{2+} in regulating glucagon secretion, with Ca^{2+} having a permissive and cAMP having an instructive role (Tengholm and Gylfe 2017). Somatostatin is one of the signals that can decrease glucagon secretion. As in β cells, somatostatin signals through the G proteins coupled to somatostatin receptors, and these G proteins inactivate adenylate cyclase. When applied to islets incubated in low glucose (1–3 mM), somatostatin decreases cAMP levels (Elliott et al. 2015), and conversely, SSTR2 antagonists increase cAMP levels (Elliott et al. 2015; Yu et al. 2019).

On the other hand, experiments detailing the α cell response to somatostatin signaling in moderately high glucose have been inconsistent. Some groups have reported a decrease of cAMP in response to 7 mM glucose, even in the presence of SSTR2 antagonists (Tengholm and Gylfe 2017; Yu et al. 2019; Vieira et al. 2007), suggesting that somatostatin has little to no role in decreasing glucagon secretion in response to elevated glucose. In contrast, work from the Piston lab demonstrated that SSTR2 antagonists increase cAMP in 11 mM glucose (Elliott et al. 2015). The two groups employed similar experimental conditions except for glucose concentration. Somatostatin, unlike glucagon, has an almost linear response to increased glucose with increasing secretion even at the highest glucose level examined, 30 mM (Gylfe 2016). Taken together, these data may support a model in which intrinsic α cell mechanisms silence glucagon secretion at glucose levels near 7 mM, but as glucose rises to and past 11 mM, SSTR signaling steps in to help silence α cells. However, when glucose passes a certain threshold, neither α-cell intrinsic methods nor δ cell input can prevent α-cell triggering, and glucagon is secreted at inappropriately high glucose levels.

Experimental evidence of somatostatin directly repressing δ-cell function is lacking, but somatostatin inhibition of δ cells likely occurs as it does in α and β cells; cAMP and Ca^{2+} are principal molecules regulating somatostatin secretion (Denwood et al. 2019).

5 Extrinsic Versus Intrinsic Control of δ-Cell Function

Mechanisms of δ-cell function control are unclear, in large part because of the low number of δ cells within islets, the intermixing of δ cells with other islet cells, and until recently, the inability to examine δ cells specifically. Recently, interest in δ cells has increased, and a few contrasting but reconcilable ideas for the means of δ-cell regulation have been proposed.

One such idea is that δ cells are triggered by β-cell action. Urocortin 3, released from β cells along with insulin, acts through the corticotropin-releasing hormone receptor 2 in δ cells to stimulate somatostatin release (van der Meulen et al. 2015; Li et al. 2018). β Cells also activate δ cells through gap junctions (Zaborska et al. 2022; Briant et al. 2018). The triggering of δ cells by β cells, along with a dampening effect of somatostatin on other secretory cells, has led to the theories that δ cells act in a feedback loop to inhibit β-cell function (van der Meulen et al. 2015), or that βcells stimulate δ cells to inhibit glucagon release as glucose rises (Briant et al. 2018).

While it is clear that electrical activity and secreted molecules from β cells trigger δ cells, β-cell signaling cannot be the sole means by which δ cells are stimulated. δ cells release somatostatin at lower glucose levels than β cells release insulin (Gylfe 2016) and, therefore, must be activated by signals outside of those originating in the β cell. Mouse δ cells release substantial amounts of somatostatin at 5 mM glucose, while β cells lag until concentrations above 10 mM glucose. What, then, controls somatostatin release outside of β-cell control? δ cells may react to inputs such as glucose very similarly to β cells. Mouse δ cells express high levels of genes—comparable to those in β cells—required for a direct response to glucose (Fig. 2), including *Glucokinase*, insulin-independent glucose transporters (*Slc2a1* and *Slc2a3,* but not *Slc2a2*, the latter of which is high in β cells), adenylate cyclase (*Adcy6*, *Adcy8*, *Adcy9*), components of the ATP-sensitive potassium channel (*Kcnj11* and *Abcc8*), and a voltage-gated calcium channel (*Kcna5*) (DiGruccio et al. 2016).

Experimental evidence from functional studies also suggests that δ cells respond directly to changes in glucose concentration. In intact islets, 10 mM glucose stimulates δ membrane depolarization, action potential firing, and somatostatin secretion, as does the application of the ATP-dependent potassium channel inhibitor

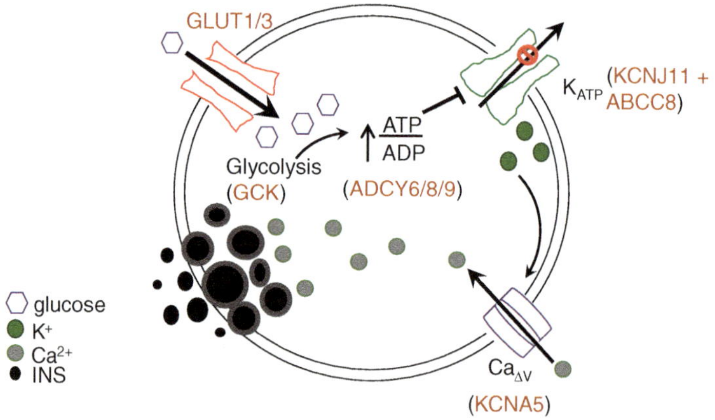

Fig. 2 RNA-seq indicates that δ cells express machinery capable of responding directly to changing glucose concentration, including glucose transporters, adenylate cyclase, components of the ATP-sensitive K^+ channel, and voltage-gated Ca^{2+} channels (DiGruccio et al. 2016). Protein names in this pathway are highlighted in burnt umber

tolbutamide, albeit not as strongly as glucose (Denwood et al. 2019). The authors interpreted this difference in the strength of the two stimuli to mean that pathways and molecules independent of the ATP-dependent potassium channel but which respond to changing glucose concentrations are essential for δ-cell activation. One of these molecules is cAMP, which also increased in response to both 10 and 20 mM glucose and to glucagon-like peptide 1 (GLP-1) (Denwood et al. 2019). cAMP invokes somatostatin release in part by releasing calcium stores from the endoplasmic reticulum (Denwood et al. 2019; Zhang et al. 2007). In δ cells, somatostatin release is not always correlated with cytoplasmic calcium levels, whereas it is with cAMP, indicating that cAMP may be the more critical messenger controlling somatostatin release in δ cells (Denwood et al. 2019).

6 δ Cell Arrangement and Structure Imply a Vital Role Within the Islet

As mentioned above, islet architecture varies from species to species. In rats, rabbits, and mice, a mantle of α, δ, and PP cells surround a core of β cells (Shen et al. 2012; Levetan and Pierce 2013; Kim et al. 2009). In primates, including baboons (Quinn et al. 2012; Guardado Mendoza et al. 2015), macaques (Kim et al. 2009), and humans (Cabrera et al. 2006; Levetan and Pierce 2013), endocrine cells within the islet are more intermixed, while pig and dog islet structures lie between those of rodents and primates (Kim et al. 2009). However, whether primate endocrine cell organization is genuinely different from that of other species is controversial. It has been reported that smaller islets in both human and mouse pancreas have β-cell cores and mantels composed of other endocrine cells. In comparison, larger islets have multiple small clusters of β cells, each surrounded by α, δ, and PP cells repetitively throughout the islet (Bonner-Weir et al. 2015). However, the three-dimensional reconstruction of human islets contradicts this claim (Dybala and Hara 2019) and suggests the model in which there is an intermixing of cell types in human islets is correct.

An islet arrangement in which all endocrine cell types are comingled and the percentage of β cells is lower might mandate a higher intensity and influence for intra-islet communications compared to an islet composed primarily of segregated β cells. These differences could indicate a more sizeable effect on β cells by α and δ cells in primate islets compared to species with more intermixed islets. However, it has been demonstrated that δ cells in both human and mouse islets have axon-like projections that can stretch from 5 to 30 μm and move during the response to glucose (Arrojo et al. 2019). This latter finding supports an influence of δ cells on both α and β cells that are important enough to require a distinct morphology among islet endocrine cells so that each δ cell can touch multiple α and β cells, which have an average diameter of around 9–10 μm (Sasaki et al. 2021; Marchetti et al. 2017).

7 δ Cells and Diabetes

The apparent importance of δ cells in islet regulation—the control of both α- and β-cell functions and the shift in morphology in mouse islets with long projections when they are located at the periphery of the islet—implies that they have a role in the development of diabetes, but is there any evidence for it?

Several organisms demonstrate a decline in δ cell numbers either before diabetes onset or during diabetes progression. In mice with a homozygous mutation in the leptin receptor (*Lepr*$^{db/db}$ mice), δ cell number initially increases at 8–10 weeks of age compared to controls with a peak at 14 weeks of age. However, total δ-cell volume is reduced in *Lepr*$^{db/db}$ mice by 20 weeks of age, a time point where the percentage of δ cells within the islet was substantially reduced compared to controls (Leiter et al. 1979). In a spontaneous baboon model of diabetes, δ cell loss precedes changes in glucose homeostasis and β cell mass (Guardado Mendoza et al. 2015). This decline in δ-cell number could indicate that δ-cell loss contributes to the onset or progression of diabetes; however, it could also support a role for δ cells in replenishing a failing β-cell population through transdifferentiation (Carril Pardo et al. 2022; Chera et al. 2014).

In humans, comparing the β:δ cell ratio in adult controls and donors with T2D also suggests that δ cells contribute to diabetes onset or progression. In adult humans without diabetes, the β:δ cell ratio is unchanged with age. On the other hand, this ratio is lower in younger adults with T2D (Fig. 3a) (Kaestner et al. 2019). This data indicates young adults with T2D have more δ cells relative to β cells than healthy controls. This difference in the β:δ cell ratio narrows with age. Together, this data suggests that a high number of δ cells, potentially coupled with increased somatostatin secretion, could be a contributory factor for adults with early-onset T2D. In healthy controls and in donors with T2D, the α:δ cell ratio decreased with time (Fig. 3b).

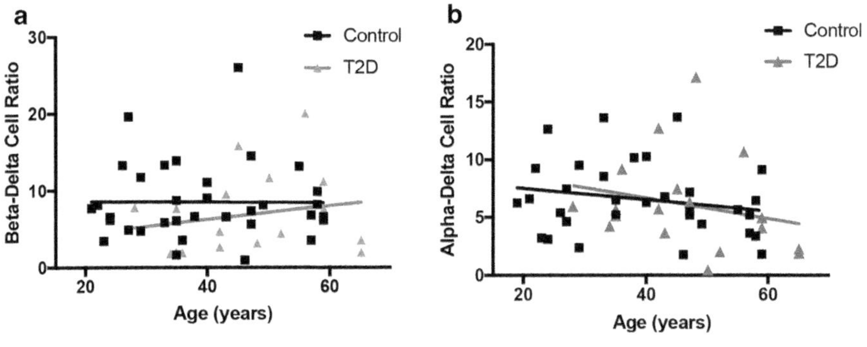

Fig. 3 Islet δ-cell ratios compared to age in control or T2D donors. (**a**) β:δ cell ratio in human donors with or without T2D by age. (**b**) α:δ cell ratio in human donors with or without T2D by age. Data for this figure was generated from the Human Pancreas Analysis Program mass cytometry dataset (Kaestner et al. 2019)

8 Conclusions

Although the somatostatin-expressing nature of the pancreatic δ cell was first described in 1975 (Goldsmith et al. 1975; Orci et al. 1975), the study of pancreatic δ cells, their role in glucose homeostasis, and how their function is controlled is still nascent, having been overshadowed by the more abundant α and β cells and the obvious importance of insulin in blood glucose homeostasis. Recent technological advancements, especially single-cell assays, have permitted research into the rarer δ cell in the context of the intact islet and otherwise. Currently, what stimulus is the primary regulator of δ-cell function, and what role the δ cell plays in the development of diabetes, is unclear. However, the changing numbers and ratios of δ cells in many species with the onset or progression of diabetes, and the ability of somatostatin analogs to reversibly induce diabetes in some circumstances point to the value of more δ-cell-focused research. Furthermore, somatostatin antagonists have the potential to be used, if only in conjunction with targeted delivery, to silence inappropriately activated α cells in the face of high glucose, a contributing factor to T2D (D'Alessio 2011).

Acknowledgments This chapter was supported by NIDDK grant (1R01DK110183 to MLG). This chapter used data acquired from the Human Pancreas Analysis Program (HPAP-RRID:SCR_016202) Database (https://hpap.pmacs.upenn.edu), a Human Islet Research Network (RRID:SCR_014393) consortium (UC4-DK-112217, U01-DK-123594, UC4-DK-112232, and U01-DK-123716) to generate Fig. 3.

Compliance with Ethical Standards Maria L. Golson declare no conflict of interest. This chapter does not contain any studies with any human participants or animals performed by any of the authors.

References

Adriaenssens AE, Svendsen B, Lam BY, Yeo GS, Holst JJ, Reimann F, Gribble FM (2016) Transcriptomic profiling of pancreatic alpha, beta and delta cell populations identifies delta cells as a principal target for ghrelin in mouse islets. Diabetologia 59(10):2156–2165

Ampofo E, Nalbach L, Menger MD, Laschke MW (2020) Regulatory mechanisms of somatostatin expression. Int J Mol Sci 21(11)

Arrojo EDR, Jacob S, Garcia-Prieto CF, Zheng X, Fukuda M, Nhu HTT, Stelmashenko O, Pecanha FLM, Rodriguez-Diaz R, Bushong E, Deerinck T, Phan S, Ali Y, Leibiger I, Chua M, Boudier T, Song SH, Graf M, Augustine GJ, Ellisman MH, Berggren PO (2019) Structural basis for delta cell paracrine regulation in pancreatic islets. Nat Commun 10(1):3700

Benuck M, Marks N (1976) Differences in the degradation of hypothalamic releasing factors by rat and human serum. Life Sci 19(8):1271–1276

Bo Q, Yang F, Li Y, Meng X, Zhang H, Zhou Y, Ling S, Sun D, Lv P, Liu L, Shi P, Tian C (2022) Structural insights into the activation of somatostatin receptor 2 by cyclic SST analogues. Cell Discov 8(1):47

Bonner-Weir S, Sullivan BA, Weir GC (2015) Human islet morphology revisited: human and rodent islets are not so different after all. J Histochem Cytochem 63(8):604–612

Briant LJB, Reinbothe TM, Spiliotis I, Miranda C, Rodriguez B, Rorsman P (2018) delta-cells and beta-cells are electrically coupled and regulate alpha-cell activity via somatostatin. J Physiol 596(2):197–215

Cabrera O, Berman DM, Kenyon NS, Ricordi C, Berggren PO, Caicedo A (2006) The unique cytoarchitecture of human pancreatic islets has implications for islet cell function. Proc Natl Acad Sci U S A 103(7):2334–2339

Cappellani D, Urbani C, Sardella C, Scattina I, Marconcini G, Lupi I, Manetti L, Marcocci C, Bogazzi F (2019) Diabetes mellitus induced by somatostatin analogue therapy is not permanent in acromegalic patients. Endocrinol Diabetes Metab 2(1):e00033

Carril Pardo CA, Massoz L, Dupont MA, Bergemann D, Bourdouxhe J, Lavergne A, Tarifeno-Saldivia E, Helker CS, Stainier DY, Peers B, Voz MM, Manfroid I (2022) A delta-cell subpopulation with a pro-beta-cell identity contributes to efficient age-independent recovery in a zebrafish model of diabetes. Elife 11

Chera S, Baronnier D, Ghila L, Cigliola V, Jensen JN, Gu G, Furuyama K, Thorel F, Gribble FM, Reimann F, Herrera PL (2014) Diabetes recovery by age-dependent conversion of pancreatic delta-cells into insulin producers. Nature 514(7523):503–507

Cigliola V, Ghila L, Thorel F, van Gurp L, Baronnier D, Oropeza D, Gupta S, Miyatsuka T, Kaneto H, Magnuson MA, Osipovich AB, Sander M, Wright CEV, Thomas MK, Furuyama K, Chera S, Herrera PL (2018) Pancreatic islet-autonomous insulin and smoothened-mediated signalling modulate identity changes of glucagon(+) alpha-cells. Nat Cell Biol 20(11):1267–1277

Colturi TJ, Unger RH, Feldman M (1984) Role of circulating somatostatin in regulation of gastric acid secretion, gastrin release, and islet cell function. Studies in healthy subjects and duodenal ulcer patients. J Clin Invest 74(2):417–423

D'Alessio D (2011) The role of dysregulated glucagon secretion in type 2 diabetes. Diabetes Obes Metab 13(Suppl 1):126–132

Denwood G, Tarasov A, Salehi A, Vergari E, Ramracheya R, Takahashi H, Nikolaev VO, Seino S, Gribble F, Reimann F, Rorsman P, Zhang Q (2019) Glucose stimulates somatostatin secretion in pancreatic delta-cells by cAMP-dependent intracellular Ca(2+) release. J Gen Physiol 151(9):1094–1115

DiGruccio MR, Mawla AM, Donaldson CJ, Noguchi GM, Vaughan J, Cowing-Zitron C, van der Meulen T, Huising MO (2016) Comprehensive alpha, beta and delta cell transcriptomes reveal that ghrelin selectively activates delta cells and promotes somatostatin release from pancreatic islets. Mol Metab 5(7):449–458

Dorrell C, Abraham SL, Lanxon-Cookson KM, Canaday PS, Streeter PR, Grompe M (2008) Isolation of major pancreatic cell types and long-term culture-initiating cells using novel human surface markers. Stem Cell Res 1(3):183–194

Duttaroy A, Zimliki CL, Gautam D, Cui Y, Mears D, Wess J (2004) Muscarinic stimulation of pancreatic insulin and glucagon release is abolished in m3 muscarinic acetylcholine receptor-deficient mice. Diabetes 53(7):1714–1720

Dybala MP, Hara M (2019) Heterogeneity of the human pancreatic islet. Diabetes 68(6):1230–1239

Elliott AD, Ustione A, Piston DW (2015) Somatostatin and insulin mediate glucose-inhibited glucagon secretion in the pancreatic alpha-cell by lowering cAMP. Am J Physiol Endocrinol Metab 308(2):E130–E143

Finan B, Capozzi ME, Campbell JE (2020) Repositioning glucagon action in the physiology and pharmacology of diabetes. Diabetes 69(4):532–541

Francis BH, Baskin DG, Saunders DR, Ensinck JW (1990) Distribution of somatostatin-14 and somatostatin-28 gastrointestinal-pancreatic cells of rats and humans. Gastroenterology 99(5):1283–1291

Goldsmith PC, Rose JC, Arimura A, Ganong WF (1975) Ultrastructural localization of somatostatin in pancreatic islets of the rat. Endocrinology 97(4):1061–1064

Gomes-Porras M, Cardenas-Salas J, Alvarez-Escola C (2020) Somatostatin analogs in clinical practice: a review. Int J Mol Sci 21(5)

Guardado Mendoza R, Perego C, Finzi G, La Rosa S, Capella C, Jimenez-Ceja LM, Velloso LA, Saad MJ, Sessa F, Bertuzzi F, Moretti S, Dick EJ Jr, Davalli AM, Folli F (2015) Delta cell death in the islet of Langerhans and the progression from normal glucose tolerance to type 2 diabetes in non-human primates (baboon, Papio hamadryas). Diabetologia 58(8):1814–1826
Gylfe E (2016) Glucose control of glucagon secretion—'There's a brand-new gimmick every year'. Ups J Med Sci 121(2):120–132
Haris B, Saraswathi S, Hussain K (2020) Somatostatin analogues for the treatment of hyperinsulinaemic hypoglycaemia. Ther Adv Endocrinol Metab 11:2042018820965068
Hauge-Evans AC, King AJ, Carmignac D, Richardson CC, Robinson IC, Low MJ, Christie MR, Persaud SJ, Jones PM (2009) Somatostatin secreted by islet delta-cells fulfills multiple roles as a paracrine regulator of islet function. Diabetes 58(2):403–411
Hauge-Evans AC, Bowe J, Franklin ZJ, Hassan Z, Jones PM (2015) Inhibitory effect of somatostatin on insulin secretion is not mediated via the CNS. J Endocrinol 225(1):19–26
Hernandez C, Simo-Servat O, Simo R (2014) Somatostatin and diabetic retinopathy: current concepts and new therapeutic perspectives. Endocrine 46(2):209–214
Holz GG (2004) Epac: A new cAMP-binding protein in support of glucagon-like peptide-1 receptor-mediated signal transduction in the pancreatic beta-cell. Diabetes 53(1):5–13
Honey RN, Weir GC (1980) Acetylcholine stimulates insulin, glucagon, and somatostatin release in the perfused chicken pancreas. Endocrinology 107(4):1065–1068
Huising MO, van der Meulen T, Huang JL, Pourhosseinzadeh MS, Noguchi GM (2018) The difference delta-cells make in glucose control. Physiology (Bethesda) 33(6):403–411
Hunyady B, Hipkin RW, Schonbrunn A, Mezey E (1997) Immunohistochemical localization of somatostatin receptor SST2A in the rat pancreas. Endocrinology 138(6):2632–2635
Kaestner KH, Powers AC, Naji A, Consortium H, Atkinson MA (2019) NIH initiative to improve understanding of the pancreas, islet, and autoimmunity in type 1 diabetes: the human pancreas analysis program (HPAP). Diabetes 68(7):1394–1402
Kaihara KA, Dickson LM, Jacobson DA, Tamarina N, Roe MW, Philipson LH, Wicksteed B (2013) beta-Cell-specific protein kinase A activation enhances the efficiency of glucose control by increasing acute-phase insulin secretion. Diabetes 62(5):1527–1536
Kailey B, van de Bunt M, Cheley S, Johnson PR, MacDonald PE, Gloyn AL, Rorsman P, Braun M (2012) SSTR2 is the functionally dominant somatostatin receptor in human pancreatic beta- and alpha-cells. Am J Physiol Endocrinol Metab 303(9):E1107–E1116
Kalwat MA, Cobb MH (2017) Mechanisms of the amplifying pathway of insulin secretion in the beta cell. Pharmacol Ther 179:17–30
Kim A, Miller K, Jo J, Kilimnik G, Wojcik P, Hara M (2009) Islet architecture: A comparative study. Islets 1(2):129–136
Kumar U, Sasi R, Suresh S, Patel A, Thangaraju M, Metrakos P, Patel SC, Patel YC (1999) Subtype-selective expression of the five somatostatin receptors (hSSTR1-5) in human pancreatic islet cells: a quantitative double-label immunohistochemical analysis. Diabetes 48(1):77–85
Leiter EH, Gapp DA, Eppig JJ, Coleman DL (1979) Ultrastructural and morphometric studies of delta cells in pancreatic islets from C57BL/Ks diabetes mice. Diabetologia 17(5):297–309
Levetan CS, Pierce SM (2013) Distinctions between the islets of mice and men: implications for new therapies for type 1 and 2 diabetes. Endocr Pract 19(2):301–312
Li N, Yang Z, Li Q, Yu Z, Chen X, Li JC, Li B, Ning SL, Cui M, Sun JP, Yu X (2018) Ablation of somatostatin cells leads to impaired pancreatic islet function and neonatal death in rodents. Cell Death Dis 9(6):682
Loud FB, Holst JJ, Egense E, Petersen B, Christiansen J (1985) Is somatostatin a humoral regulator of the endocrine pancreas and gastric acid secretion in man? Gut 26(5):445–449
Marchetti P, Bugliani M, De Tata V, Suleiman M, Marselli L (2017) Pancreatic beta cell identity in humans and the role of type 2 diabetes. Front Cell Dev Biol 5:55
Orci L, Baetens D, Dubois MP, Rufener C (1975) Evidence for the D-cell of the pancreas secreting somatostatin. Horm Metab Res 7(5):400–402

Patel YC, Greenwood MT, Warszynska A, Panetta R, Srikant CB (1994) All five cloned human somatostatin receptors (hSSTR1-5) are functionally coupled to adenylyl cyclase. Biochem Biophys Res Commun 198(2):605–612
Quinn AR, Blanco CL, Perego C, Finzi G, La Rosa S, Capella C, Guardado-Mendoza R, Casiraghi F, Gastaldelli A, Johnson M, Dick EJ Jr, Folli F (2012) The ontogeny of the endocrine pancreas in the fetal/newborn baboon. J Endocrinol 214(3):289–299
Rodriguez-Diaz R, Dando R, Jacques-Silva MC, Fachado A, Molina J, Abdulreda MH, Ricordi C, Roper SD, Berggren PO, Caicedo A (2011) Alpha cells secrete acetylcholine as a non-neuronal paracrine signal priming beta cell function in humans. Nat Med 17(7):888–892
Rohrer SP, Schaeffer JM (2000) Identification and characterization of subtype selective somatostatin receptor agonists. J Physiol Paris 94(3-4):211–215
Rorsman P, Huising MO (2018) The somatostatin-secreting pancreatic delta-cell in health and disease. Nat Rev Endocrinol 14(7):404–414
Sasaki H, Saisho Y, Inaishi J, Watanabe Y, Tsuchiya T, Makio M, Sato M, Nishikawa M, Kitago M, Yamada T, Itoh H (2021) Reduced beta cell number rather than size is a major contributor to beta cell loss in type 2 diabetes. Diabetologia 64(8):1816–1821
Saunders DC, Brissova M, Phillips N, Shrestha S, Walker JT, Aramandla R, Poffenberger G, Flaherty DK, Weller KP, Pelletier J, Cooper T, Goff MT, Virostko J, Shostak A, Dean ED, Greiner DL, Shultz LD, Prasad N, Levy SE, Carnahan RH, Dai C, Sevigny J, Powers AC (2019) Ectonucleoside triphosphate diphosphohydrolase-3 antibody targets adult human pancreatic beta cells for in vitro and in vivo analysis. Cell Metab 29(3):745–754. e4
Shamsi BH, Chatoo M, Xu XK, Xu X, Chen XQ (2021) Versatile functions of somatostatin and somatostatin receptors in the gastrointestinal system. Front Endocrinol (Lausanne) 12:652363
Shen XX, Li HL, Pan L, Hong J, Xiao J, Hermansen K, Jeppesen PB, Li GW (2012) Glucotoxicity and alpha cell dysfunction: involvement of the PI3K/Akt pathway in glucose-induced insulin resistance in rat islets and clonal alphaTC1-6 cells. Endocr Res 37(1):12–24
Singh V, Brendel MD, Zacharias S, Mergler S, Jahr H, Wiedenmann B, Bretzel RG, Plockinger U, Strowski MZ (2007) Characterization of somatostatin receptor subtype-specific regulation of insulin and glucagon secretion: an in vitro study on isolated human pancreatic islets. J Clin Endocrinol Metab 92(2):673–680
Steiner DJ, Kim A, Miller K, Hara M (2010) Pancreatic islet plasticity: interspecies comparison of islet architecture and composition. Islets 2(3):135–145
Strowski MZ, Parmar RM, Blake AD, Schaeffer JM (2000) Somatostatin inhibits insulin and glucagon secretion via two receptors subtypes: an in vitro study of pancreatic islets from somatostatin receptor 2 knockout mice. Endocrinology 141(1):111–117
Strowski MZ, Kohler M, Chen HY, Trumbauer ME, Li Z, Szalkowski D, Gopal-Truter S, Fisher JK, Schaeffer JM, Blake AD, Zhang BB, Wilkinson HA (2003) Somatostatin receptor subtype 5 regulates insulin secretion and glucose homeostasis. Mol Endocrinol 17(1):93–106
Stueven AK, Kayser A, Wetz C, Amthauer H, Wree A, Tacke F, Wiedenmann B, Roderburg C, Jann H (2019) Somatostatin analogues in the treatment of neuroendocrine tumors: past, present and future. Int J Mol Sci 20(12)
Taniguchi H, He M, Wu P, Kim S, Paik R, Sugino K, Kvitsiani D, Fu Y, Lu J, Lin Y, Miyoshi G, Shima Y, Fishell G, Nelson SB, Huang ZJ (2011) A resource of Cre driver lines for genetic targeting of GABAergic neurons in cerebral cortex. Neuron 71(6):995–1013
Tengholm A, Gylfe E (2017) cAMP signalling in insulin and glucagon secretion. Diabetes Obes Metab 19(Suppl 1):42–53
Tyagi A, Pugazhenthi S (2021) Targeting insulin resistance to treat cognitive dysfunction. Mol Neurobiol 58(6):2672–2691
van der Meulen T, Donaldson CJ, Caceres E, Hunter AE, Cowing-Zitron C, Pound LD, Adams MW, Zembrzycki A, Grove KL, Huising MO (2015) Urocortin3 mediates somatostatin-dependent negative feedback control of insulin secretion. Nat Med 21(7):769–776
Vieira E, Salehi A, Gylfe E (2007) Glucose inhibits glucagon secretion by a direct effect on mouse pancreatic alpha cells. Diabetologia 50(2):370–379

Wang YJ, Golson ML, Schug J, Traum D, Liu C, Vivek K, Dorrell C, Naji A, Powers AC, Chang KM, Grompe M, Kaestner KH (2016a) Single-cell mass cytometry analysis of the human endocrine pancreas. Cell Metab 24(4):616–626

Wang YJ, Schug J, Won KJ, Liu C, Naji A, Avrahami D, Golson ML, Kaestner KH (2016b) Single-cell transcriptomics of the human endocrine pancreas. Diabetes 65(10):3028–3038

Yu Q, Shuai H, Ahooghalandari P, Gylfe E, Tengholm A (2019) Glucose controls glucagon secretion by directly modulating cAMP in alpha cells. Diabetologia 62(7):1212–1224

Zaborska KE, Jordan KL, Thorson AS, Dadi PK, Schaub CM, Nakhe AY, Dickerson MT, Lynch JC, Weiss AJ, Dobson JR, Jacobson DA (2022) Liraglutide increases islet Ca(2+) oscillation frequency and insulin secretion by activating hyperpolarization-activated cyclic nucleotide-gated channels. Diabetes Obes Metab 24(9):1741–1752

Zeyda T, Diehl N, Paylor R, Brennan MB, Hochgeschwender U (2001) Impairment in motor learning of somatostatin null mutant mice. Brain Res 906(1-2):107–114

Zhang Q, Bengtsson M, Partridge C, Salehi A, Braun M, Cox R, Eliasson L, Johnson PR, Renstrom E, Schneider T, Berggren PO, Gopel S, Ashcroft FM, Rorsman P (2007) R-type Ca(2+)-channel-evoked CICR regulates glucose-induced somatostatin secretion. Nat Cell Biol 9(4):453–460

Zhang J, McKenna LB, Bogue CW, Kaestner KH (2014) The diabetes gene Hhex maintains delta-cell differentiation and islet function. Genes Dev 28(8):829–834

Sweet Spot Regulation of Maternal Metabolic Health and Nutrition on β-Cell Mass in the Offspring

Grace Chung, Alicia Wong, Tracy Her, and Emilyn U. Alejandro

Abstract Maternal nutrition and metabolic health status during pregnancy are critical factors that shape the life-long health trajectory of offspring. Altered nutrition during specific times of development in utero can lead to functional changes in tissues such as the pancreatic β-cells, predisposing those tissues to metabolic diseases and Type 2 diabetes that manifest later in life. This chapter will focus on the role of pregnancy complications with altered nutrition during gestation in the maladaptive programming of β-cell mass and function in the offspring.

Keywords DOHaD pancreas development · Islets · Diabetes · Maternal health · Low protein · Maternal obesity · Gestational diabetes

Abbreviations

CR	Calorie restriction
E	Embryonic day
FGR	Fetal growth restriction
GCK	Glucokinase
GD	Gestational day
GDM	Gestational diabetes mellitus
GK	Goto Kakizaki
GLP-1	Glucagon-like peptide
GSIS	Glucose-stimulated insulin secretion

Grace Chung, Alicia Wong and Tracy Her contributed equally with all other contributors.

G. Chung · A. Wong · T. Her · E. U. Alejandro (✉)
Department of Integrative Biology and Physiology, University of Minnesota Medical School, Minneapolis, MN, USA
e-mail: chung491@umn.edu; wongx615@umn.edu; herxx397@umn.edu; ealejand@umn.edu

M. L. Golson (ed.), *Molecular and Cell Biology of Pancreas Development, Function and Regeneration*, Advances in Anatomy, Embryology and Cell Biology 239, https://doi.org/10.1007/978-3-031-62232-8_7

HFD	High-fat diet
HFS	High fat and sucrose
HNF	Hepatocyte nuclear factor 4α
IL-1β	Interleukin 1 β
IL-6	Interleukin 6
LP	Low protein
mTOR	Mammalian target of rapamycin
NCD	Normal chow diet
P	Postnatal day
Pdx1	Pancreatic and duodenal homeobox 1
PPARs	Peroxisome proliferator-activated receptors
ROS	Reactive oxygen species
STZ	Streptozotocin
T2D	Type 2 diabetes
TNFα	Tumor necrosis factor α
WSD	Western-style diet

1 Introduction

Suboptimal pregnancy conditions can have far-reaching consequences on the health of the offspring across its lifespan. Rigorous human epidemiological studies in humans and experimental animal studies show that fetal nutrient factors in utero confer significant susceptibility risk for obesity and Type 2 diabetes (T2D) later in life. For example, maternal malnutrition during human pregnancy causes fetal growth restriction, a risk factor for metabolic disease and T2D later in life (Ravelli et al. 1976; de Rooij et al. 2006a, b) (Fig. 1). Maternal obesity or a Western diet during pregnancy also contributes to obesity and T2D in adult offspring (Fig. 1). Moreover, gestational diabetes mellitus (GDM) during pregnancy predisposes offspring to obesity and T2D in adulthood. Such adverse outcomes highlight the importance of optimal nutritional status during pregnancy in regulating organ growth and programming multiple metabolic cells, such as the pancreatic β-cell and insulin-sensitive tissues. The in utero environment strongly influences the numerous signaling pathways required for the embryonic and postnatal development of the endocrine pancreas. Based on human and rodent studies, β-cells are exquisitely sensitive to the fetal nutrient environment. Thus, the in utero environment could ultimately impact adult β-cell mass (Stanger et al. 2007) and the subsequent risk for T2D later in life (Stanger et al. 2007; Alejandro et al. 2015).

The etiology of T2D is complex. Obesity is a major risk factor for insulin resistance and T2D. In the state of insulin resistance, T2D develops when there is a significant loss of functional β-cell mass and the amount of insulin secreted is no longer adequate to meet the demands of peripheral tissues, resulting in hyperglycemia. Thus, a deficit in β-cell mass and insulin secretion determines whether an individual

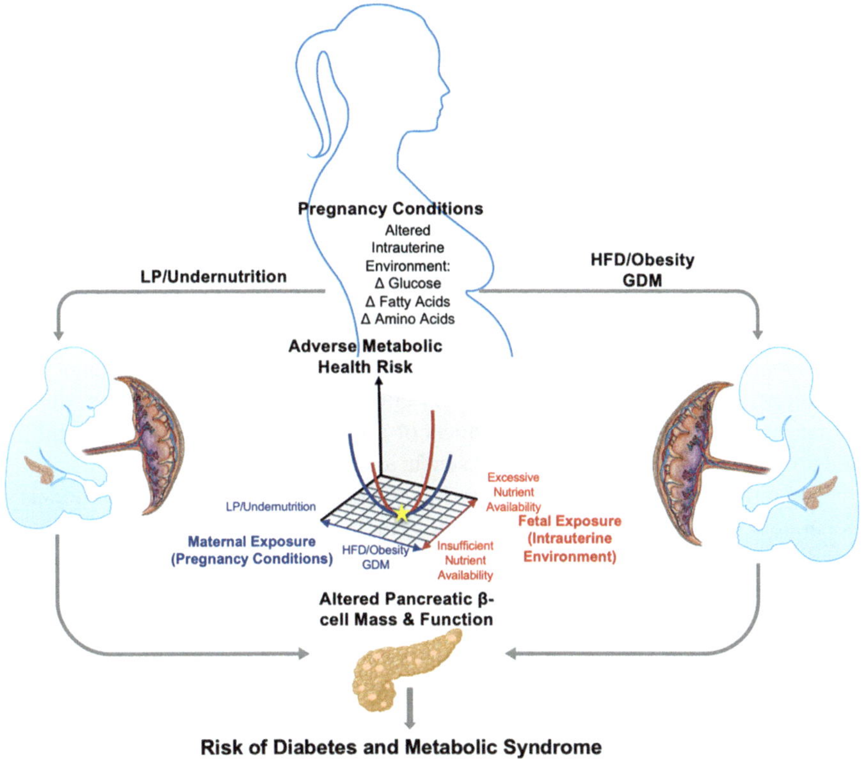

Fig. 1 Sweet spot of nutritional programming of fetal metabolic health. Pregnancy conditions such as gestational diabetes (GDM), obesity, and undernutrition (LP, low protein) perturb the intrauterine environment and alter the availability and transport of nutrients to the fetus via the placenta. These changes are associated with, among other factors, genetic, epigenetic, and inflammatory alterations that program fetal metabolic health parameters such as β-cell mass and function and glucose homeostasis in the short-term, long-term, and across generations. Maternal nutrient intake and availability of nutrients to the fetus in the intrauterine environment converge at a "sweet spot" that allows for optimal fetal growth and development

will develop T2D (Ferrannini et al. 2005). A plateau of β-cell mass, or the total population of insulin-producing cells in adult humans, is likely achieved during the first decade of postnatal life (Gregg et al. 2012). In adulthood, β-cell number is maintained by meager self-replication of existing β-cells (Dor et al. 2004). While human data support this concept, elevated β-cell mass attained during early stages may provide some measure of protection from T2D (Alejandro et al. 2015).

During pregnancy, the fetus is entirely dependent on the maternal nutrient supply. Thus, the quality and adequacy of intrauterine conditions are primarily determined by maternal nutrition and metabolic health status. Alterations in this nutrient supply can profoundly impact the structure and function of crucial fetal organs, including the fetal β-cells, which are sensitive to glucose, amino acid, and lipid levels during development (Aerts and Van Assche 2006; Devlieger et al. 2008).

Thus, the mother's nutrient availability and dietary behavior during pregnancy affect the nutrients sensed or transported into the fetus and may contribute significantly to the fetal programming of β-cell mass and function.

Preclinical models have afforded us the advantages of detailed in vivo and in vitro mechanistic approaches to study metabolic and physiological outcomes in longitudinal studies of offspring cohorts in various models. An important lesson we have learned is that the duration and types of insult matter, and that these parameters should be carefully considered when interpreting data. For example, the amount of protein restriction or HFD and duration of treatment (throughout gestation or for the last week of pregnancy, and cessation with birth or through the lactation period) may dictate the severity of the effects. The resulting metabolic syndrome may manifest sooner when the insult period is prolonged (i.e., throughout pregnancy and lactation). The study of maternal diet restriction at different time points has demonstrated the individual and distinct effects of a low-protein diet (LP) and identified the critical "developmental window" sensitive to LP levels.

Although the scientific community has made a great effort to create models to accurately mimic human pathophysiology, each model has limitations. For example, in maternal obesity models, the challenge comes partly from the timing of obesity induction and the diet composition (e.g., 60% vs. 35% kcal with sucrose) inducing obesity. Given that GDM is a multifactorial disease, no single environmental or genetic factor can recapitulate the disease. Thus, the generation of a model that mimics the disease is complex. However, to move forward, developing effective animal models is essential for understanding the short- and long-term consequences of developmental programming of maternal pregnancy complications, providing insights into potential treatments and preventative measures to combat obesity and T2D.

In this chapter, we will highlight fetal programming paradigms that model pregnancy complications such as maternal low-protein diet, maternal obesity, and GDM that give rise to obesity and T2D. We will discuss available data on the impacts of these models on pancreatic β-cell development and function and postulate mechanistic links between pregnancy complications and the offspring's metabolic health trajectory.

2 Maternal Low Protein/Undernutrition Exposures

The placenta is a dynamic organ that can adapt to changes in the maternal and fetal environment; under suboptimal conditions, however, it fails to adapt completely (Sandovici et al. 2012). Placental insufficiency leads to fetal growth restriction, a condition in which the lack of nutrients from the placenta can cause the fetus to fail to achieve its maximum potential growth and size (Mohan et al. 2018). Maternal calorie restriction (CR) and low-protein diet (LP) animal models provide evidence that nutrients are integral in early fetal pancreatic β-cell development (Cox et al. 2010). Amino acids play a vital role in protein synthesis for growth and development. Whether alone or in a cocktail with isoleucine and valine, leucine induces a sudden onset of increased circulating insulin levels in humans, rodents, and mammals (Yang et al. 2010). For example, Gao et al. cultured rat islets for 1 week with leucine and found increased ATP content and calcium concentrations, leading to more insulin secretion. Leucine has been studied extensively; it activates the mammalian target of rapamycin (mTOR) (Yang et al. 2010; Rachdi et al. 2012), which is critical for pancreas development. Mice lacking pancreatic epithelial mTOR show pancreas hypoplasia (Elghazi et al. 2017). Leucine promotes protein synthesis and β-cell proliferation by activating mTOR signaling (Elghazi et al. 2017; Yang et al. 2004). As a nutrient-sensing kinase, mTOR integrates nutrients and growth factor signaling pathways. Therefore, alterations in mTOR signaling are involved in fetal β-cell programming. For example, mTOR is decreased in the islets of offspring from LP dams (King et al. 2019; Alejandro et al. 2014). The reduction in β-cell area induced by maternal LP is rescued by increasing mTOR specifically in the offspring β-cells, but only if mTOR elevation occurs during pregnancy (Alejandro et al. 2014). In many malnourished animal models, such as CR, fetal growth restriction (FGR) leads to long-term changes in β-cell mass and function (Mohan et al. 2018; Yang et al. 2004; Limesand et al. 2005) (summarized in Table 1). FGR is associated with increased susceptibility to metabolic diseases in adulthood, such as obesity and T2D (Park et al. 2008).

Table 1 Effect of maternal low-protein diet during pregnancy on offspring β-cells

Model	Diet composition	Effect on β-cell mass	Effect on β-cell function
Pregnant C57BL/6 exposed to LP throughout pregnancy, LP0.5 (Alejandro et al. 2014)	Isocaloric to chow and 9% protein	Neonates ↓ β-cell fraction Adult 12 weeks Males • ↑ β-cell mass Females • ↓ β-cell mass	Neonates ↓ Insulin levels Adult 12 weeks ↑ Insulin sensitivity ↓ Insulin secretion Glucose intolerance
Pregnant C57BL/6 exposed to LP at e12.5, LP12.5 (Akhaphong et al. 2021)	Isocaloric to chow and 9% protein	Neonates Normal β-cell fraction Adult 48–51 weeks Males • ↑ β-cell mass Females • No change	Neonates No change in insulin levels Adult 6 weeks Males/Females Normal glucose tolerance Adult 48–51weeks Male Glucose intolerance Normal insulin sensitivity Female Glucose intolerance Insulin resistance

(continued)

Table 1 (continued)

Model	Diet composition	Effect on β-cell mass	Effect on β-cell function
Pregnant Wistar rats fed LP throughout gestation (RC) and/or lactation (RR) (Rodriguez-Trejo et al. 2012)	Isocaloric to chow 6% protein	PND 7/21 RC/RR ↓ β-cell mass	PND 7/21 RC/RR ↑ Insulin levels Glucose intolerance
Pregnant Sprague-Dawley rats, fed LP on G11 throughout pregnancy (Yuan et al. 2011)	Isocaloric to chow 9% protein	Adulthood ↓ β-cell mass	Adulthood Glucose intolerance Insulin resistance
Shropshire twin-pregnant sheep, fed LP during the last 6 weeks of gestation (Kongsted et al. 2014)	50% of daily requirements for energy and protein for normal twin pregnancy	–	Adult 6 months LP + NCD Normal glucose tolerance LP+ High-fat diet Females ↓ Insulin area Insufficient insulin secretion
Columbia-Ram-bouillet crossbred ewes singleton pregnancy, exposed to 40 °C from gestational day 37 ±3 (Limesand et al. 2005)	–	Fetal ↓ β-cell mass	Fetal ↓ Insulin expression
Pregnant Rhesus Macaques, LP diet fed 1 month before breeding and throughout pregnancy (Kirigiti et al. 2020)	50% protein-restricted diet	6 months old ↑ β- to α-cell ratio No distinct changes	6 months old Hyperinsulinemia
Pregnant Baboons, from 30 days of gestation through lactation, were fed 70% restricted feed of control (Choi et al. 2011)	70% of feed eaten by control	–	3.5 years ↑ Fasting glucose ↑ Fasting insulin Glucose intolerance
Dutch hunger winter families (Ravelli et al. 1976; de Rooij et al. 2006a, b)		–	↑ 2-h glucose levels after a glucose load ↓ after a glucose load

LP0.5, low-protein throughout pregnancy, gestation day 0.5 to birth; LP12.5, low-protein during the last week of pregnancy, gestation day 12.5 to birth; NCD, normal chow diet; ↑, increased or enhanced; ↓, decreased or impaired

2.1 Maternal LP Rodent Models

We and others have shown that maternal nutritional stress, such as a low-protein (LP) diet, total calorie restriction, or uteroplacental insufficiency during gestation, induces FGR. FGR is a critical risk factor that predisposes offspring to T2D (Sandovici et al. 2013) by causing long-term consequences in β-cell mass and function (Alejandro et al. 2014; Fernandez-Twinn and Ozanne 2010; Snoeck et al. 1990; Hoet 1991; Simmons et al. 2001; Akhaphong et al. 2018) and abnormal offspring metabolic health in adulthood. Robust rodent data show adverse effects on offspring β-cell mass and function when dams are fed an LP diet during pregnancy and lactation. However, the results are mixed depending on the length of LP exposure. Murine offspring exposed to LP throughout pregnancy (LP0.5) show reduced β-cell mass at birth, partly due to impaired proliferation and enhanced apoptosis, followed by glucose intolerance in the adult offspring (Alejandro et al. 2015; Dumortier et al. 2007; Heltemes et al. 2010; Petrik et al. 1999a; Rodríguez-Trejo et al. 2012). Interestingly, feeding pregnant Balb/c females an isocaloric LP diet had no impact on female or male offspring β-cell mass at birth, but 14-day-old female offspring demonstrated increased β-cell mass (Cox et al. 2010). As adults, these females exhibited normalization of β-cell area (Alejandro et al. 2014). Despite the lack of changes in total β-cell area, the adult offspring of both C57BL/6J or Balb/c LP dams develop glucose intolerance even with improved insulin sensitivity (Cox et al. 2010; Alejandro et al. 2014; Su et al. 2016). These data highlight that a defect in β-cell function, namely their ability to secrete insulin in response to high-glucose exposure, causes this glucose intolerance (Alejandro et al. 2014). Under stress conditions, the offspring islets show additional altered responses, such as a reduced ability to recover β-cell mass after streptozotocin (STZ)-induced β-cell death (Cox et al. 2010). A more prolonged low-protein exposure through weaning resulted in a significantly decreased β-cell mass at birth (Cox et al. 2010; King et al. 2019; Su et al. 2016; Petrik et al. 1999b). The offspring of C57BL/6J dams exposed to a LP diet throughout pregnancy and through the end of the lactation period also display glucose intolerance, decreased β-cell mass, and a defect in insulin secretion in adulthood (Su et al. 2016).

Interestingly, a shorter exposure window, with LP only during the last week of pregnancy (LP12.5), results in normal β-cell area at birth and normal glucose tolerance in young adult C57BL/6 mice (Alejandro et al. 2020a). Notwithstanding a defect in β-cell area at birth, both sexes developed glucose intolerance with aging (48 weeks of age) (Alejandro et al. 2020a). Although not seen in males, female mice eventually developed insulin resistance with age. Interestingly, at 48 weeks of age, male offspring showed an increase in β-cell mass, whereas female offspring displayed no significant changes compared to their control counterparts (Alejandro et al. 2020a). However, this result contrasts with previous studies in Wistar rats, in which LP12.5 offspring show reduced β-cell area, partly due to impaired β-cell proliferation (Dumortier et al. 2007). The offspring of Wistar or Sprague-Dawley rat LP dams also display reduced β-cell mass and perturbed insulin secretion capacity (Mohan et al. 2018; Rodríguez-Trejo et al. 2012; Chamson-Reig et al. 2006; Haase et al. 2013). Increasing leucine and arginine increases insulin secretion in LP fetal islets in vitro (Dahri et al. 1991). However, it is important to note that insulin levels are severely diminished compared to controls, suggesting LP induces permanent functional damage in utero (Dahri et al. 1991).

2.2 Paternal LP Diet

There is strong evidence that maternal metabolic and nutritional status during pregnancy is closely linked to the metabolic health trajectory of offspring later in life. To our knowledge, the effect of paternal LP diet on pancreatic β-cell mass and function is unexplored. In the few available rodent studies, adult offspring of sires fed LP developed glucose intolerance compared to progeny of normal chow diet (NCD)-fed sires. In contrast to maternal LP diet-inducing FGR, paternal LP offspring display increased fetal weight and are heavier with growth progression (Watkins and Sinclair 2014; Morgan et al. 2021). Altered calcium signaling and metabolism were observed in the heart and liver of LP offspring (Watkins and Sinclair 2014). In these studies, effects on the pancreas were not tested. However, we speculate that perturbed calcium signaling would impair β-cell insulin secretion, which would cause glucose intolerance. Like maternal programming, paternal factors may shape β-cell development and function and need further research.

2.3 Sheep

Extensive studies provide similar evidence of changes in the β-cell when pregnant ewes are fed a late-gestation LP diet. In twin pregnancies, at 6 months of age, LP offspring develop insulin resistance compared to those born from mothers on an NCD diet. When placed on a healthy diet, young lambs of LP dams sustain normal glucose tolerance by hypersecreting insulin. However, glucose tolerance was perturbed when these offspring were challenged with a high-fat and high-sugar diet. Adult females display decreased insulin area and insufficient insulin secretion, which may lead to the development of T2D due to a lack of compensation in high-glucose environments (Kongsted et al. 2014).

Additionally, FGR can be induced in progeny by exposing pregnant ewes to elevated temperatures. One study in singleton-pregnant sheep showed that overall pancreas weight decreased by about 60%. In these offspring, fetal pancreatic β-cell mass was decreased by 76%, and insulin expression was reduced significantly per pancreas. Interestingly, glucagon content remained unaltered and only changed based on pancreas size (Limesand et al. 2005). As previously mentioned, LP can lead to placental insufficiency characterized by reduced placenta size and glucose and amino acid transport (Mohan et al. 2018). Placental insufficiency that causes FGR in fetal sheep results in decreased β-cell mass, insulin secretion, and increased β-cell proliferation (Table 1) (Limesand et al. 2007).

2.4 Non-human Primates

Few studies on the effects of a low-protein diet during gestation have been carried out in non-human primates, thus limiting β-cell mass data. Interestingly, non-human primate studies indicate no distinct changes in β-cell mass when dams were administered an LP diet. Kirigiti et al. demonstrated that rhesus macaques given a 50%

protein-restricted diet 1 month before breeding and throughout pregnancy had offspring that at 6 months developed hyperinsulinemia and exhibited a trend toward an increased β- to α-cell ratio (Kirigiti et al. 2020). In one study, baboons were limited to 70% of calories usually consumed starting from gestational day 30 and throughout pregnancy. This calorie restriction recapitulated studies reducing caloric intake in rodents, which resulted in the offspring developing fasting hyperinsulinemia. After 3.5 years, weaned offspring were given a NCD, yet they still manifested increased fasting glucose, fasting insulin, and glucose intolerance (Choi et al. 2011; Palou et al. 2010).

2.5 *Humans*

In humans, malnourishment during pregnancy leads to IUGR and FGR, conditions in which babies are born with a low birth weight. FGR has been associated with various metabolic, renal, and cardiovascular diseases later in adulthood (Mohan et al. 2018). However, few studies examine β-cell mass or function in FGR offspring. However, strong correlations have been observed between low birth weight and the risk for obesity and T2D (Mohan et al. 2018). An FGR diagnosis is strongly associated with altered insulin sensitivity in children and adults (Newsome et al. 2003). Still, it is important to note that impaired insulin sensitivity is not sufficient to cause T2D by itself. Historically, observational studies around the world have shown that individuals exposed to intrauterine malnutrition due to famine are at a higher risk of developing hyperglycemia (Vipin et al. 2022). For example, people exposed to the Dutch famine of 1944–1945 during early gestation had an increased risk for heart diseases and T2D (Roseboom et al. 2003). Evidence from this study (de Rooij et al. 2006a) suggests that mid-gestation famine exposure results in glucose intolerance mediated by reduced insulin secretion.

2.6 *Potential Mechanisms of the Effects of Fetal Changes in Nutrition on β-Cell Development and Function*

In most mammalian species, both paternal and maternal LP diets negatively impact glucose metabolism, in part by altering progeny β-cell mass and function. In utero, many factors, including replication and apoptosis, play a crucial role in fetal pancreas development (Limesand et al. 2005). The fetal period is a critical stage in which cellular replication, growth, and dedifferentiation are all sensitive to nutrient availability. Therefore, intrauterine changes can have long-lasting effects on the metabolic health of offspring (Rando and Simmons 2015). A sufficient supply of nutrients during intrauterine growth is vital for developing a full complement of β-cells; these cells exhibit reduced proliferation with age. Various mechanisms regarding the impact of maternal LP diet on fetal β-cell development have been proposed. Growth regulators such as mTOR have a crucial role in preventing FGR (King et al. 2019; Alejandro et al. 2014). Genetic models have demonstrated that mTOR plays a critical role in pancreas development (Elghazi et al. 2017) and β-cell maturation (Jaafar et al. 2019), maintenance (Blandino-Rosano et al. 2017), and function (Alejandro et al. 2017). When mTOR expression is decreased, nutrients are not correctly sensed by the fetus, leading to decreased β-cell mass (Alejandro et al. 2014).

A reduction in pancreatic and duodenal homeobox 1 (Pdx1), a crucial transcription factor for pancreas development and β-cell development, maturation, and identity maintenance (Gao et al. 2014; Stoffers et al. 1997), has also been reported in islets of LP offspring (Alejandro et al. 2014; Abuzgaia et al. 2015). Ozanne et al. demonstrated the importance of epigenetic regulation through hepatocyte nuclear factor (HNF)4α, a key transcription factor involved in pancreatic β-cell differentiation and glucose homeostasis (Ozanne et al. 2011). Igf2 is also reduced in LP islets (Sparre et al. 2003).

Small non-coding miRNAs regulate many developmental and physiological processes across diverse cells and tissues, including β-cells. Several of these miRNAs have been implicated in the fetal programming of β-cell dysfunction (miRNA 375, 342, 199a–3p, 15b, 7) by LP diet (Alejandro et al. 2014; Su et al. 2016; Dumortier et al. 2014). These miRNAs target multiple key proteins relevant to β-cell health, such as mTOR and OGT. For example, blocking miRNA-199a-3p and miRNA 342 action was sufficient to restore both mTOR levels and insulin secretion (Alejandro et al. 2014).

Inflammation, particularly macrophages, has also been implicated in β-cell programming by FGR, albeit in a non-LP model (Root et al. 2022). A LP diet also induces increased expression of tumor necrosis factor α (TNFα), interleukin 1 β (IL-1β), and interleukin 6 (IL-6) in rat placenta (Qi et al. 2022). In one human study, women with IUGR and placental insufficiency displayed increased TNFα and pro-inflammatory cytokines (Bartha et al. 2003). Root et al. demonstrated that macrophage depletion improves the β-cell-mass deficit induced by FGR (Root et al. 2022). Since LP diet and placental insufficiency both increase inflammation within the placenta, future studies should investigate the impact of placental inflammation on β-cell development.

3 Maternal Obesity Exposures

Obesity is the most common chronic health condition affecting women of reproductive age and is associated with an increased risk of insulin resistance, obesity, and metabolic syndrome in adult offspring (ACOG 2021; Fall et al. 1998; Dearden et al. 2020; Eriksson et al. 2015; Schack-Nielsen et al. 2010; Laitinen et al. 2001; Eriksson et al. 2014; Patel et al. 2016). Maternal obesity is correlated with alterations in nutrient supply, which is a determinant of both placental function and offspring metabolic health outcomes (Lucas 1998; Lucas et al. 1999; Parrettini et al. 2020; Kelly et al. 2020). One such nutrient is fatty acids. Fatty acid availability and transport are critical for fetal development, affecting placental function and growth and fetal growth (Duttaroy and Basak 2020). Modeling maternal obesity to effectively recapitulate human pathophysiology has been challenging, and studies have used different paradigms or timelines to implement HFD: (1) HFD before and during pregnancy and NCD after birth; (2) HFD before and during pregnancy and during lactation; (3) HFD before and NCD during pregnancy and lactation. The effects of these models are summarized in Table 2. Models involving prolonged HFD throughout gestation recapitulate sustained elevated body weight and adiposity throughout pregnancy and/or lactation. On the other hand, a pre-pregnancy-specific HFD regimen seeks to model maternal obesity while excluding confounding effects of maternal diet. Maternal HFD is associated with molecular and epigenetic changes that affect offspring behavior and predisposition to obesity and metabolic disease (Gawlińska et al. 2021; Saengnipanthkul et al. 2021; Peng et al. 2021; Pang et al. 2021).

3.1 Pre-pregnancy Maternal High-Fat Diet

3.1.1 Fetal and Neonatal

A common way to induce maternal obesity and to study its effects on fetal development is by subjecting a mother to HFD feeding before and throughout pregnancy. However, an ongoing objective in the field is to elucidate a critical window of HFD feeding that most closely mimics the human pathophysiology of maternal obesity. The effect of maternal nutrition during gestation is well-reviewed (Gawlińska et al. 2021; Geraghty et al. 2015; Lowensohn et al. 2016; Govoni et al. 2019; Kunz and King 2007). To circumvent the direct impact of elevated nutrition during gestation, HFD can be fed to the mother pre-pregnancy to induce obesity, followed by a normal diet composition during gestation. Much of the information on the effect of pre-pregnancy-specific HFD-induced maternal obesity on offspring β-cell mass and function early in life comes from Qiao et al.'s 2019 study in the C57BL/6J mouse (Qiao et al. 2019). Hallmarks of maternal obesity, such as increased maternal adiposity, body weight, and circulating insulin throughout pregnancy can be recapitulated by HFD feeding (60% kcal from fat) 3 months before pregnancy, followed by a transition to NCD feeding during gestation (Qiao et al. 2019). Regardless of sex, third-trimester fetuses (embryonic day 18.5) and neonates (postnatal day 1) from dams fed this diet regimen exhibit decreased blood glucose levels and increased circulating insulin, and consistent with this physiology, increased insulin signaling and increased β-cell area. Both maternal HFD feeding and maternal obesity are associated with increased fatty acid supply to the fetus, suggesting a potential

Table 2 Effect of maternal and paternal HFD on offspring glucose homeostasis

Animal/time on HFD	Age	Diet fat composition	Glucose homeostasis	β-Cell mass	α-Cell mass	Mechanism
Macaque 2–5 years (Comstock et al. 2012)	GD130, 6 months (pre-wean), 7 months (pre-wean) 13 months (postnatal)	31.8% kcal	6 months ↔ insulin secretion 7 months↑ of insulin resistance	↔ β-Cell mass (GD130, 13 months) ↔ β-Cell number (13 months)	↓ α-cell mass (GD130) ↔ α-cell number (13 months)	
Macaque 2+ years (Elsakr et al. 2019)	13 months	36% kcal + once daily calorically dense treat	↔ Glucose tolerance ↑ Insulin secretion	↔ β-Cell mass ↑ β-Cell/ α-cell ratio	↓ α-Cell mass	
Sheep Pre-pregnancy through Trimester 1 (Ford et al. 2009)	GD75	150% of NRC recommendation	↑ Blood glucose ↑ Insulin ↑ Cortisol	↑ β-Cell area ↑ β-Cell number	↔ α-Cell number	↑ β-Cell proliferation, ↑ Glucose exposure, ↑ cortisol
C57/BL6 Pre-pregnancy (Qiao et al. 2019)	E18.5, P1	60% kcal	↑ Serum insulin ↑GSIS (20 mM)	↑ β-Cell area		↑ Maternal fatty acid supply
C57BL/6J 4 weeks pre-pregnancy and throughout gestation (Akhaphong et al. 2021)	P1, 16–26 weeks	60% kcal	Neonate: ↓ blood glucose, ↓ β-cell fraction adult ♂ ↔ Glucose tolerance (NCD), ↓ glucose tolerance (HFD) Adult ♀ ↔ glucose tolerance (NCD) ↑ insulin resistance (NCD) ↑ insulin secretion (HFD)	↓ β-Cell fraction (p1) ↔ (females, adult)	↔ α-Cell fraction (p1) ↔ (females, adult)	

C57BL/6J 8 weeks pre-pregnancy and throughout gestation and lactation (Bringhenti et al. 2013)	P0, P10, 12 weeks	49% kcal (32% lard, 17% other lipids)	12 weeks: ↓ glucose tolerance ↑ insulin	β-Cell mass/ pancreas mass ↓ (P0) ↔ (P10) ↔ (12 weeks)		
C57BL/6 3 weeks pre-pregnancy through gestation and lactation (Zheng et al. 2020)	Weanling	60% kcal	↓ Glucose tolerance ↑ Insulin resistance	↑ β-Cell area, mass		↑ β-Cell proliferation, ↔ apoptosis
C57BL/6 6 weeks pre first-pregnancy through gestation and lactation of second pregnancy (Nicholas et al. 2020)	7 and 8 weeks	20% hard lard (and condensed milk)	♀: ↑ Plasma insulin, GSIS; ↓ blood glucose ♂: ↔ GSIS All: ↔ glucose tolerance			♀: ↑ Mitochondrial respiration
C57BL/6 F0: 8 weeks pre-pregnancy through gestation and lactation (Graus-Nunes et al. 2015)	3 months	49% kcal	F1: ↓ glucose tolerance, ↑ fasting glucose, F1, F2: ↑ insulin	F1, F2: ↑ β-cell mass	F1, F2: ↑ α-cell mass	

(continued)

Table 2 (continued)

Animal/time on HFD	Age	Diet fat composition	Glucose homeostasis	β-Cell mass	α-Cell mass	Mechanism
Wistar Rat Trimester 1, 2, 3, or 1–3 (Cerf et al. 2005, 2009)	P1	40% kcal	Trimester1: ↓ blood glucose Trimester 2: ↓ blood glucose Trimester 1–3: ↑ blood glucose, ↓ insulin release All: ↔ circulating insulin	Trimester 1,3: ↓ β-cell number Trimester 1–3: ↓ β-cell volume, number	Trimester 1,3: ↑ α-cell size Trimester 1–3: ↑ α-cell volume, number, size	Impaired PDX1 and GCK immunoreactivity
Wistar Rat Trimester 1, 2, 3 or 1–3 (Cerf et al. 2007)	Weanling	40% kcal	Trimester 1, 2, or 3: ↑ blood glucose All: ↓ body weight, ↓ insulin	Trimester 1, 2, or 3: ↓ β-cell volume, number, size	Trimester 1, 2, or 3: ↓ α-cell size	↓ GCK mRNA and reactivity
Wistar Rat Gestation and lactation, gestation only, lactation only (Cerf and Louw 2010)	Weanling	40% kcal	Gestation and Lactation, Lactation only: ↑ blood glucose ↓ glucose tolerance Gestation only: Mild ↓ glucose tolerance All: ↓ circulating insulin			Impaired β-cell function due to ↓GCK expression and immunoreactivity
Wistar Rat *F0*: 6 weeks pre-pregnancy through gestation and lactation. *F1,2*: NCD after weaning (Schellong et al. 2020)	3 weeks and 3–5 months	34% kcal	F1, F2: ↑ predisposition to diabesity (↑ blood glucose, ↑ insulin resistance)			

Sprague-Dawley Rat *F0*: 8 weeks pre-pregnancy through gestation and lactation *F1, F2*: NCD after Weaning (Huang et al. 2017)	F2: 3 and 12 weeks	45% kcal	↓ Glucose tolerance, ↓ serum insulin			↓ β-Cell proliferation
Sprague-Dawley Rat ♂: from 4 weeks of age ♀: NCD (Ng et al. 2010)	13 and 14 weeks	40.7%, 43% kcal	↓ GSIS ↓ glucose tolerance			Δ Gene methylation in offspring islets

β-cells, α-cells, and potential mechanisms. Offspring weaned onto NCD unless otherwise stated. GD, gestation day; E, embryonic day; P, postnatal day; NCD, normal chow diet; HFD, high-fat diet; GSIS, glucose-stimulated insulin secretion; ↑, increased/enhanced; ↓, decreased/impaired; ↔, no change relative to control group mechanistic link between fatty acid availability and β-cell development (Qiao et al. 2015, 2019; Diaz et al. 2015). Indeed, Qiao et al. demonstrate that hindering lipolysis in the placenta through ablation of adipose triglyceride lipase diminishes enhanced β-cell development associated with maternal HFD in pre-pregnancy (Qiao et al. 2019).

3.1.2 Later Life and Metabolic Challenge

While Qiao et al. beautifully demonstrated the effect of maternal pre-gestational HFD on β-cell development in neonates, it is unclear how these β-cells respond to metabolic stress post-weaning. Liu et al.'s study in B6/129/SvEv mice demonstrate the differential effects of pre-gestational HFD and gestational HFD on offspring β-cell adaptation to postnatal HFD challenge (Liu et al. 2020). Unsurprisingly, in utero exposure to HFD has a detrimental impact on offspring β-cell function. Male offspring from dams subjected to HFD during gestation display increased β-cell size in response to HFD feeding but are insulin resistant and glucose intolerant. In contrast, male offspring from HFD-fed dams that transition to an NCD 9 weeks before pregnancy are protected from these effects, mainly through mitigation of β-cell proliferation and insulin granulation defects.

Interestingly, female offspring display similar glucose and insulin tolerance trends but have no histological changes in the islet (Liu et al. 2020). These results suggest a sexual dimorphism in regulating β-cell compensation to dietary stress. Nonetheless, this evidence indicates that diet composition during gestation, in addition to maternal obesity, has significant effects on offspring β-cell development and function.

3.2 High-Fat Diet Before and During Gestation

3.2.1 Fetal and Neonatal

Rodents

Much knowledge about the effect of gestational HFD comes from rodent studies in mice and rats. A typical model involves pre-treating a dam with HFD before pregnancy and extending this treatment through gestation only or through gestation and the lactation period combined. Interestingly, male offspring subjected to gestational HFD exhibit no differences in body weight at birth but display increased body weight and blood glucose later in neonatal development (Akhaphong et al. 2021; Bringhenti et al. 2013). When assessing male or combined-sex offspring, neonatal offspring from dams fed HFD show a decreased β-cell fraction or decreased β-cell mass, despite unchanged *Pdx1* expression or cell division (Akhaphong et al. 2021; Bringhenti et al. 2013). While these differences normalize with age, offspring from

HFD-fed mothers display a greater increase in β-cell mass between birth and 10 days of age and between birth to 3 months of age, suggesting in utero exposure to HFD affects ex utero β-cell expansion (Bringhenti et al. 2013).

The effect of maternal HFD feeding on offspring metabolism depends on the gestation period and perhaps species. Rat studies have results both contrasting and contradicting mouse studies. Wistar rat offspring exposed to maternal HFD during the first or second trimester of gestation exhibit reduced blood glucose levels. Conversely, HFD feeding throughout gestation results in hyperglycemic offspring (Cerf et al. 2005). In contrast to the mouse model, exposure to HFD only during the first trimester of pregnancy results in reduced offspring birth weight (Cerf et al. 2005). Consistent with mouse studies, despite no differences in circulating insulin concentrations, offspring of rat dams fed HFD throughout pregnancy have reduced β-cell volume. In contrast to mice, though, α-cell volume is increased (Cerf et al. 2005). While exposure to HFD in trimesters 1 and 3 results in reduced β-cell number and increased α-cell size in rats, the effect of HFD throughout the entire duration of gestation had the most pronounced adverse impact on β-cell development and function (Cerf et al. 2005). Mechanistically, HFD-exposed offspring display impaired insulin release due to reduced PDX1 and glucokinase (Cerf et al. 2009).

Gestational period-specific effects of the intrauterine environment are not entirely surprising. Studies in individuals born during the Dutch famine linked adverse long-term health outcomes of offspring to maternal undernutrition, with effects differing based on the timing of gestational hunger exposure (Roseboom et al. 2003, 2006). Duttaroy and Basak comprehensively reviewed the critical role of fatty acids in placental nutrient transport throughout pregnancy on fetal metabolic outcomes. Long-chain fatty acids play a role in essential angiogenesis events during placentation and, consequently, fetal growth. In contrast, third-trimester transport of long-chain polyunsaturated fatty acids from the mother to the fetus is a mediator of fetal brain development (Duttaroy and Basak 2020). Maternal diet during pregnancy affects the transport of fatty acids to the fetus. Given the role of fatty acid availability in β-cell expansion in the fetus, HFD feeding during parts of or all of gestation leads to varying metabolic consequences in the offspring (Duttaroy and Basak 2020; Qiao et al. 2015).

Larger Mammals

A concern with rodent models of pregnancy is the effect of short gestational length, fetal number, and placentation differences on the effective recapitulation of human pregnancy. Sheep and non-human primates are larger mammals with extended gestational periods commonly used to model nutritional effects on offspring. The consequences of pre-pregnancy and gestational maternal HFD feeding in these larger mammals are mixed.

When sheep dams are fed HFD before conception, and through the first trimester of gestation, much like the observations in the murine model of pre-pregnancy-specific

HFD and in contrast to gestational HFD, fetuses exhibit elevated circulating glucose and insulin and increased β-cell area, with no changes in α-cell area (Ford et al. 2009). Rodent studies support increased maternal glucose production and turnover and increased placental glucose transport during pregnancy in HFD-fed dams (Jones et al. 2009; Song et al. 2017; Leturque et al. 1987). This result is mechanistically similar to the HFD sheep experiments in which increased β-cell area in the offspring of HFD-fed dams' results from elevated β-cell proliferation brought on by increased exposure to glucose and cortisol (Ford et al. 2009).

In contrast to sheep studies, data in the non-human primate is mainly null. Comstock et al. found that juvenile offspring of HFD-fed female macaques show no differences in physiological markers of metabolism, such as body weight, blood glucose, or circulating insulin (Comstock et al. 2012). Additionally, these progeny show normal islet morphology, neogenesis, proliferation, and apoptosis. Consistent with the lack of islet morphological differences, no changes in the transcripts of critical β-cell maintenance genes, such as *Pax6*, *Neurod1*, *Pdx1*, or *Nkx6.1*, are observed. Interestingly, however, maternal HFD affects α-cell mass and plasticity, with suppression of some transcripts in α-cells, like *Foxa2* and *Irx2* (Comstock et al. 2012). Similarly, offspring of macaques fed a Western Style Diet (WSD; high-fat diet feed and one calorically dense snack per day during gestation and lactation) display no differences in β-cell mass or size but have reduced α-cell mass (Elsakr et al. 2019). Interestingly, however, offspring weaned onto NCD exhibit higher glucose-stimulated insulin secretion (GSIS), suggesting maternal diet alters islet function (Elsakr et al. 2019). In both human and rodent models, higher α-cell content leads to improved GSIS due to paracrine effects (reviewed by Moede et al. (Moede et al. 2020)). The role of glucagon-like peptide (GLP-1) in the potentiation of insulin secretion was first described in rats and has since been validated in other model systems (Schmidt et al. 1985; Fridolf et al. 1991; Orskov et al. 1988). It is also a known regulator of islet function in both murine and human islets (de Souza et al. 2020). Additionally, evidence in the mouse model suggests that terminally differentiated α-cells have the potential to convert to β-cells following β-cell injury and loss (Thorel et al. 2010). During gestation, maternal β- and α-cell mass expand, followed by β-cell apoptosis postpartum (Rodriguez et al. 2021). In mice, α-cells can convert to β-cells postpartum (Rodriguez et al. 2021). The link between α-cell availability and plasticity for compensation during metabolic stress warrants further exploration.

3.2.2 Weanlings

Studies on the effect of maternal HFD feeding during gestation on offspring specifically at weaning are sparse and come from rodents. Cerf and colleagues performed several experiments in Wistar rats suggesting the effects of maternal HFD during gestation are trimester-specific (Cerf et al. 2007). Contrary to studies in neonatal offspring from dams maintained on HFD for the entire duration of gestation, weanlings from dams fed HFD during specific trimesters display metabolic defects associated with compromised β-cell development and function. Exposure to HFD during each trimester alone, but not together, is associated with reduced body weight and hypoinsulinemia, accompanied by elevated blood glucose and reduced β-cell volume, number, and size in offspring. Maintenance of HFD for the entire duration of gestation has no effects on β- or α-cell development, suggesting compensation much like that observed in other gestational dietary manipulations such as low protein diet (Cerf et al. 2007).

In mice, data on maternal HFD feeding in male weanlings comes from one study by Zheng and colleagues (Zheng et al. 2020). They demonstrate that maternal HFD feeding throughout gestation and lactation is associated with increased islet size and density in offspring at weaning. β-cell area and β-cell mass are increased and mechanistically tied to increased proliferation without changes in apoptosis. Male offspring are glucose intolerant and insulin resistant (Zheng et al. 2020). However, it is unclear whether the metabolic syndrome displayed by these mice is associated solely with peripheral tissue insulin resistance and secondary β-cell compensation or also with primary defects in β-cell function. The study did not assess female offspring, and as maternal HFD is known to affect milk composition, the potential compounding effect of HFD during lactation should be evaluated (Carlson et al. 2022).

3.2.3 Long-Term Effects on β-Cells

In mice, the long-term metabolic effects of maternal HFD during gestation are well-studied, but the specific impact on β-cell mass later in life is less clear. Bringhenti et al. described no changes in β-cell mass in adult (12 weeks) male offspring of mouse dams exposed to HFD before and during pregnancy, but these offspring display greater β-cell mass expansion between birth (when β-cell mass is reduced) and p10 and between birth and 12 weeks of age. 12-week-old offspring also display elevated glucose and circulating insulin but no differences in body weight (Bringhenti et al. 2013). Interestingly, Akhaphong et al. demonstrated comparable glucose tolerance in male offspring from dams exposed to HFD during gestation and controls, although differences in fasting body weight observed at 16 weeks of age were normalized by 20 weeks of age. When these mice were challenged to HFD for 8 weeks, 26-week-old gestationally HFD-exposed male mice exhibited a trend toward glucose intolerance. 50-week-old mice challenged to HFD for 32 weeks showed blunted GSIS, suggesting incomplete compensation over time (Akhaphong et al. 2021). In contrast, adult female offspring of HFD-fed dams displayed insulin resistance at 20–22 weeks of age but showed no differences in β- or α-cell mass or size, suggesting sexual dimorphism of islet intrauterine programming (Akhaphong et al. 2021).

3.2.4 Combinatorial Effect of Maternal HFD During Lactation

In rats, Cerf and colleagues demonstrated evidence for combinatorial effects of maintaining maternal HFD through gestation and lactation (Cerf and Louw 2010; Cerf et al. 2006). Exposure to HFD during both gestation and lactation or only during lactation results in sustained increased blood glucose concentration and overt glucose intolerance, compared to HFD exposure only during gestation. Treatment with HFD during gestation and a subsequent return to NCD during lactation results in glucose intolerance in weanlings but to a milder extent than those exposed until weaning, suggesting the importance of early postnatal diet on β-cell function. Nonetheless, exposure to HFD during gestation only or during both gestation and lactation is associated with reduced GLUT2 and glucokinase expression in the pancreas, although immunoreactivity was increased and decreased, respectively. Reduced glucokinase expression is a potential mechanistic explanation for glucose intolerance in weanlings of HFD-exposed mothers (Cerf and Louw 2010; Cerf et al. 2006). A murine model of maternal HFD also suggests an additive detrimental effect of maternal HFD on male offspring glucose tolerance (Masuyama and Hiramatsu 2014).

3.2.5 Sexual Dimorphism in the Programming of Offspring Pancreatic Islets

Sex differences in metabolic programming play a role in determining offspring susceptibility to metabolic disease and may account for differences in incidence between males and females later in life (Dearden et al. 2018). A study by Nicholas et al. in 7–8-week-old male and female offspring of murine dams fed HFD during gestation and lactation indicates sex differences in the ability to adapt to hypernutrient conditions postnatally (Nicholas et al. 2020). While glucose tolerance is similar between the sexes, insulin secretion, and mitochondrial respiration are enhanced in females, and susceptibility to apoptosis is reduced compared to male littermates. Conversely, islets of male offspring display reduced stimulus-secretion coupling, due to reduced calcium ion channel expression and diminished docked insulin granules (Nicholas et al. 2020). Altogether, these data suggest that males are more susceptible to the deleterious effects of maternal HFD programming at the islet level. A contrasting study by Akhaphong et al. also demonstrate sexual dimorphism in intrauterine islet programming; however, they observed glucose intolerance in female offspring of HFD-fed dams even without postnatal HFD challenge, suggesting females are more susceptible to adverse metabolic programming under maternal HFD (Akhaphong et al. 2021). However, the setup of the two studies varied in that Akhaphong et al. kept their dams on HFD only through birth, while Nicholas et al. kept dams on HFD through lactation. Thus, the effects of milk composition and access to HFD feed on β-cell development before weaning warrants further investigation.

3.2.6 Intergenerational Effects

Few studies examine intergenerational impacts on β-cell mass development and function, and those existing implement rodent models. In Wistar rats, Huang et al. assessed the effect of F0 maternal HFD feeding on NCD-fed F2 offspring. They found impaired glucose tolerance and β-cell proliferation, along with decreased serum insulin and expression of PDX1 and NEUROD1, suggesting reduced β-cell plasticity and function (Huang et al. 2017). Similarly, Schellong et al. observed that maternal high-calorie, high-fat diet feeding increases the risk of obesity and diabetes in NCD-fed F1 and F2 generations (Schellong et al. 2020). In the mouse, both the F1 and F2 progeny of HFD-fed dams exhibit islet hypertrophy, with increased α- and β-cell mass and an increased β-cell to α-cell ratio. Both F1 and F2 offspring display reduced PDX1 immunoreactivity.

Interestingly, while F1 progeny of HFD-fed dams are glucose intolerant when evaluated at 3 months of age, no glucose intolerance is apparent in F2 offspring, likely due to transiently effective compensation brought on by hyperinsulinemia resulting from increased β-cell mass (Graus-Nunes et al. 2015). These offspring, however, display higher HOMA-IR than F2 offspring of dams fed NCD, implying glucose intolerance will develop at a later age (Graus-Nunes et al. 2015). These results suggest that maternal HFD can transgenerationally affect β-cell development and function.

3.2.7 Paternal High-Fat Diet

Fetal programming is often studied in the context of the intrauterine environment. Therefore, research on the paternal HFD effect on offspring β-cell health is sparse. Studies in mice indicate paternal obesity and HFD can affect sperm at the DNA, RNA, and epigenetic levels (Duale et al. 2014; Wang et al. 2022; Claycombe-Larson et al. 2020; Terashima et al. 2015). In sperm, miRNA and histone methylation status can affect nutrient transport and inflammation and affects offspring in a sexually dimorphic manner (Claycombe-Larson et al. 2020). In Sprague-Dawley rats, female offspring of HFD-fed fathers have impaired glucose tolerance and decreased insulin sensitivity over time. In islets of offspring, differential expression of genes involved in energy usage, cellular transport, and cell signaling was observed, along with hypomethylation of Il13ra2, which encodes a subunit of the interleukin 13 receptor complex and is a signal mediator in the Jak-Stat pathway (Ng et al. 2010). This result suggests a role of paternal HFD in offspring epigenome programming. Zhang and colleagues have aptly summarized the effects of paternal environmental exposures, such as diet, on various sperm RNAs and the effects on offspring metabolic phenotype (Zhang et al. 2019a). The mechanisms of programming and specific effects on offspring islet development warrant further study.

3.2.8 Potential Mechanisms Described

Fatty acids are essential in placental implantation and fetal development (Duttaroy and Basak 2020). In mice, maternal consumption of HFD is associated with an increase in fatty acid transport in the placenta mediated by increased expression of fatty acid transporter 6 and FABP3 (Diaz et al. 2015). Maternal HFD is also linked with β-cell hyperplasia and associated with increased placental lipoprotein lipase (LPL) and decreased SIRT1 expression (Qiao et al. 2015; Reynolds et al. 2015). Chronic HFD feeding results in β-cell mass expansion (Hull et al. 2005).

While insulin cannot cross the blood–placenta barrier, it may indirectly impact nutrient flux by regulating placental function. In contrast to insulin, glucose is readily transported from mother to fetus. Placentas of Sprague-Dawley rat dams fed HFD before and during gestation have increased expression of the glucose transporters GLUT1 and GLUT4, indicating increased glucose transport across the placenta (Reynolds et al. 2015). The Pedersen hypothesis suggests that increased fetal glucose uptake leads to increased fetal insulin secretion and increased fetal growth due to the nature of insulin as a growth factor (Macfarlane and Tsakalakos 1988).

Obesity alters the epigenome of gametes and offspring (Ou et al. 2019). Maternal HFD affects the methylation status of genes associated with insulin sensitivity, glucose metabolism, and a predisposition to metabolic syndrome (Pang et al. 2021; Zhang et al. 2019b; Keleher et al. 2018). Paternal consumption of HFD affects sperm quality, including DNA stability, various RNAs, and histone methylation status, which contribute to the regulation of metabolism in the offspring (Wang et al. 2022; Claycombe-Larson et al. 2020; Zhang et al. 2019a).

In rats, maternal HFD treatment from pre-gestation through gestation affects inflammation in the placenta in a sexually dimorphic manner. In the placenta of male offspring, markers of inflammation such as the cytokines IL1-β, TNF-α, and CD68 were elevated in response to maternal HFD feeding (Reynolds et al. 2015). Pantham and colleagues reviewed the effect of maternal obesity and GDM on maternal and fetal inflammation (Pantham et al. 2015). They concluded that while maternal obesity is associated with a low-grade inflammatory maternal environment, it is not consistently associated with fetal inflammation, partly due to placental adaptation. However, altered placental function may still play a role in fetal disease susceptibility (Pantham et al. 2015). Thus, elevated inflammation could provide a mechanistic link between maternal HFD and offspring β-cell defects.

4 Gestational Diabetes Mellitus Exposure

Gestational diabetes mellitus, or GDM, is broadly defined as glucose intolerance with onset or first recognition during pregnancy (American Diabetes Association 2013; Pasek and Gannon 2013). Typically characterized by maternal hyperglycemia, impaired glucose tolerance, and increased insulin resistance resulting from pancreatic β-cell dysfunction (Plows et al. 2018; Buchanan 2001; Buchanan et al. 1990; Kühl 1991; Homko et al. 2001; Buchanan and Xiang 2005), GDM is one of the most common pregnancy complications today. The worldwide prevalence of GDM is as high as 25%, depending on the study population and diagnostic criteria (Zhu and Zhang 2016; Schneider et al. 2012). Accumulating evidence robustly demonstrates that GDM is associated with short- and long-term complications for both the mother and infant, including an increased risk for both the mother and child to develop T2D later in life (Rayanagoudar et al. 2016; Chivese et al. 2019; Eades et al. 2015; Noctor and Dunne 2015; Herath et al. 2017; Coustan 2016). As such, GDM is an important risk factor to consider for the growing burden of diabetes.

Glucose plays a significant role in pancreatic β-cell development and differentiation (Filhoulaud et al. 2009; Guillemain et al. 2007). In pregnancies complicated by GDM, in which the fetus experiences an abundant supply of glucose, potential changes in fetal pancreatic islet development and function may impact glucose homeostasis and the long-term health of the offspring. In this section, we discuss data from human and preclinical animal studies investigating the effect of maternal diabetes on offspring β-cell development and function. Our primary focus is on glucose and the impact of exposure to increased maternal glucose supply during pregnancy on offspring health; however, the various risk factors for GDM, including obesity, prediabetes, parity, and hyperinsulinemia, which are often present together in women with GDM, are also critical confounding contributors to offspring metabolic health (Alejandro et al. 2020b). Therefore, hyperglycemia, hyperinsulinemia, and dyslipidemia can independently and likely synergistically impact the health of the offspring and should not be overlooked.

4.1 Human Studies

Understanding the specific impact of maternal GDM in humans is challenging because of the complex interactions between genetic and environmental factors. Nevertheless, epidemiological studies provide supporting evidence that intrauterine exposure to maternal hyperglycemia impacts glucose homeostasis and β-cell function in the offspring. For example, the pancreata of infants of diabetic mothers is characterized by hypertrophy and hyperplasia (Cardell 1953; D'Agostino and Bahn 1963). This expansion is accompanied by increased pancreatic insulin content and insulin secretion in the infant, contributing to fetal hyperinsulinemia (Steinke and Driscoll 1965; Reiher et al. 1983).

Similar findings were found in infants of mothers with GDM (Ogata et al. 1980). These effects in the fetal pancreas may result in alterations that persist beyond birth. Plagemann et al. studied glucose tolerance and insulin secretion in offspring (1–9 years) born to mothers with GDM (Plagemann et al. 1997). The prevalence of impaired glucose tolerance in these offspring increased from 11.1% at 1–4 years of age to 20.0% at 5–9 years. A prospective study focused on adolescent offspring (10–16 years) born to women with pre-gestational and GDM during pregnancy had an elevated prevalence of impaired glucose tolerance (Silverman et al. 1995). Assessment of fetal β-cell function, measured by amniotic fluid insulin concentration, revealed that excessive fetal insulin secretion in utero strongly correlated with glucose intolerance in the offspring. Similar results were found in a younger population (7–11 years) of offspring from mothers with GDM (Malcolm et al. 2006). In adult offspring (18–27 years) exposed to maternal GDM during pregnancy, the risk of prediabetes/diabetes, characterized by glucose intolerance or impaired fasting glucose, was eight times greater in the offspring from GDM mothers compared to that of the control group, and the risk of developing metabolic syndrome was four times greater (Clausen et al. 2008, 2009; Damm et al. 2016). This increased risk was associated with and possibly caused or exacerbated by a reduced insulin sensitivity and disposition (Kelstrup et al. 2013). A lower oral disposition index suggests a defect in β-cell compensation and is strongly associated with progression to T2D (Utzschneider et al. 2009). These findings were corroborated by the large Hyperglycemia and Adverse Pregnancy Outcome (HAPO) Follow-up Study (FUS). This study found that adolescent offspring (10–14 years) from mothers with untreated GDM had a higher risk of developing impaired glucose tolerance, reduced insulin sensitivity, and reduced oral disposition index compared to offspring from mothers without GDM. This elevated risk was independent of maternal and childhood BMI and family history of diabetes (Lowe et al. 2019; Brown et al. 2019; Scholtens et al. 2019).

While these studies highlight the potential influence of the intrauterine environment on the long-term health trajectory of the offspring, genetic predispositions constitute a significant risk factor for GDM and remain difficult to tease apart from other susceptibility influences. However, a study of Pima Indians addressed this confounding factor by examining offspring before and after a maternal diabetes diagnosis (Dabelea et al. 2000). They demonstrated a higher risk of diabetes in offspring born after the mother's diabetes diagnosis compared to siblings born before it, suggesting that the intrauterine diabetic environment is sufficient to impact the metabolic health of the offspring, independent of genetic risks. These human epidemiological studies establish that pregnancies complicated by GDM predispose the developing offspring to impaired glucose tolerance and diabetes.

4.2 Animal Studies

GDM is, by definition, transient, existing only during pregnancy. This short timeframe and the contribution of genetic and environmental risk factors make studying GDM in humans challenging. Preclinical animal models provide a valuable alternative. Several animal models are available for exploring the pathophysiology and the long-term health impact of maternal GDM on the offspring, primarily in mice and rats (Aerts and Van Assche 2006; Devlieger et al. 2008; Pasek and Gannon 2013). However, animal models of GDM do not fully capture the human pathophysiology of GDM and are limited by the parameters used to define GDM in the animal. Maternal diabetes or hyperglycemia can be induced by several methods, including the administration of streptozotocin (STZ), which selectively destroys the β-cells to cause a diabetogenic effect of hyperglycemia, diet manipulation, and embryo transplantation (Pasek and Gannon 2013).

4.3 In the Fetus or Neonate

Studies using animal models provide evidence that intrauterine exposure to an increased maternal glucose supply induces adaptation in the developing fetus. In STZ-induced mildly diabetic rats, the fetal pancreas responds to the increased glucose supply with β-cell hyperplasia and hypertrophy and increased β-cell mass (Holemans et al. 2003; Aerts et al. 1990; Reusens-Billen et al. 1984). Fetal pancreatic and plasma insulin content and GSIS were enhanced in these fetuses, resulting in hyperinsulinemia (Kervran et al. 1978). Indeed, islets isolated from fetuses from diabetic mothers incorporated more tritiated thymidine, an adulterated DNA base that can serve as a proliferation marker, than control islets (Reusens-Billen et al. 1984). On the other hand, fetal islets from STZ-induced severely diabetic mothers showed poorer granulation than control islets, suggesting a faster depletion of insulin stores than could be recovered (Aerts et al. 1990; Aerts and van Assche 1977). Eventually, the fetal islets became incapable of sufficient insulin secretion to maintain euglycemia, a conclusion supported by reduced pancreatic insulin content and plasma insulin (Kervran et al. 1978). Fetuses from severely diabetic mothers also demonstrated impaired insulin secretion in response to glucose, both in vivo and in vitro. Another model used continuous glucose infusion during late gestation to induce maternal hyperglycemia. Rat fetuses exposed to this regimen exhibited higher plasma glucose and insulin levels (Bihoreau et al. 1986; Ktorza et al. 1981). Neonates born to dams with GDM caused by a high-fat and sucrose diet (HFS) displayed significantly larger and more irregularly shaped islets than the control group (Agarwal et al. 2019). Together, these studies (summarized in Table 3) provide strong evidence that fetal pancreatic β-cells are sensitive to altered glucose flux in utero and respond with maladaptive structural and functional changes.

4.4 Metabolic Phenotypes of Adult Offspring

Young adult offspring exposed to maternal hyperglycemia induced by glucose infusion at the end of pregnancy exhibited impaired glucose tolerance and insulin secretion (Bihoreau et al. 1986; Gauguier et al. 1991). With age, the offspring became hyperglycemic at baseline and developed severely impaired glucose tolerance and insulin secretion. Consistent with these findings, another study in rat offspring from STZ-induced maternal diabetes found that young offspring from diabetic mothers exhibited normal glucose tolerance and insulin secretion. However, the older adult offspring exhibited significantly impaired insulin secretion in vivo and in vitro (Han et al. 2007). This dysfunction may result from the reduced activity of several key enzymes involved in glucose metabolism, including phosphofructokinase, pyruvate carboxylase, and pyruvate dehydrogenase. In an additional model of maternal GDM

Table 3 Effect of maternal GDM on offspring β-cell development and function

In the fetus or neonate					
Offspring age	GDM criteria	Method for GDM	Glucose homeostasis	β-Cell morphology and mass	β-Cell function
Rat fetus (GD 20–22) (Aerts et al. 1990)	Mildly diabetic mother Moderate hypoinsulinemia, hyperglycemia Severely diabetic mother Severe hypoinsulinemia, hyperglycemia	Single STZ injection on GD 1 (30 mg/kg BW) Single STZ injection on GD 1 (50 mg/kg BW)	↑ Glycemia ↑ Insulinemia ↑ Glycemia ↓ Insulinemia	Hypertrophy Hyperplasia ↑ β-Cell mass ↑ Pancreatic insulin Hypertrophy Hyperplasia ↓ Pancreatic insulin Severe β-cell degranulation	↑ Insulin secretion (in vivo and in vitro) ↓ Insulin secretion (in vivo and in vitro)
Rat fetus (GD 19.5–21.5) Sherman rats (Kervran et al. 1978)	Mildly diabetic mother BG 100–200 mg/dL Severely diabetic mother BG >300 mg/dL	Single STZ injection before mating (30–50 mg/kg BW)	↑ Glycemia ↑ Insulinemia ↑ Glycemia ↓ Insulinemia	↑ Mean pancreatic weight (GD 21.5) ↑ Pancreatic insulin ↓ Mean pancreatic weight (GD 20.5) ↓ Pancreatic insulin	↑ Insulin secretion (in vitro) ↓ Insulin secretion (in vivo and in vitro)
Rat fetus (GD 21.5) (Reusens-Billen et al. 1984)	Mildly diabetic mother BG >300 mg/dL (end of gestation)	Single dose of STZ on GD 1 (30 mg/kg BW)		Hypertrophy Hyperplasia ↑ Proliferation (in vivo and in vitro)	
Rat fetus (GD 21.5–23.5) Albino rats of Sherman strain (Ktorza et al. 1981)	Mild hyperglycemia (~12 nmol/L)	Continuous glucose infusion during the last 3 days of pregnancy	↑ Glycemia ↑ Insulinemia ↓ Plasma glucagon		
Rat neonate (GD 21.5) Albino rats of Sherman strain (Bihoreau et al. 1986)	Hyperglycemia Hyperinsulinemia	Continuous glucose infusion during last week of pregnancy	↑ Glycemia ↑ Insulinemia		

Table 3 (continued)

In the fetus or neonate

Offspring age	GDM criteria	Method for GDM	Glucose homeostasis	β-Cell morphology and mass	β-Cell function
Rat neonate (1-day-old) Sprague-Dawley (Agarwal et al. 2019)	↑ Maternal BW ↑ GWG Hyperglycemia Hyperinsulinemia	HFS diet for 6 weeks before mating (45% fat, 35% carbohydrate with 17% sucrose) and continued throughout gestation and suckling		↔ β-Cell fraction ↑ β-Cell size Irregular-shaped islets	

In the adult offspring

Offspring Age	GDM criteria	Method for GDM	Glucose homeostasis	β-Cell morphology and mass	β-Cell function
Rat adult offspring (P80–100) (Aerts et al. 1990)	Mildly diabetic mother Moderate hypoinsulinemia, hyperglycemia Severely diabetic mother Severe hypoinsulinemia, hyperglycemia	Single STZ injection on GD 1 (30 mg/kg BW) Single STZ injection on GD1 (50 mg/kg BW)	↔ Fasting glycemia ↑ Insulinemia ↔ Fasting glycemia ↑ Insulinemia ↓ Plasma glucagon	↔ Percentage of endocrine tissue ↔ β-Cell granulation ↔ Percentage of endocrine tissue ↔ β-Cell granulation ↑ Pale granules	↓ Insulin secretion (in vivo and in vitro) ↑ Insulin secretion (in vivo and in vitro)
Rat adult offspring (up to 15 weeks) Sprague-Dawley (Han et al. 2007)	Hyperglycemia BG 400–450 mg/dL	Single STZ injection before pregnancy (50 mg/kg BW)	↓ Glycemia up to 3 weeks (males) and 4 weeks (females) then normalized ↔ Glucose tolerance up to 15 weeks	↔ Insulin content ↔ Islet size ↔ β-cell mass	↓ Insulin secretion (females at 10 and 15 weeks; males at 15 weeks) (in vivo and in vitro)
Rat adult offspring (5 months) Albino Wistar rat (Taqui et al. 2020)	Fasting BG >130 mg/dL	Single STZ injection before pregnancy (90 mg/kg BW)	↑ Glycemia	↓ Percentage of islet area (males) ↓ β-cell mass (males) ↓ Total number of islets	

(continued)

Table 3 (continued)

In the adult offspring					
Offspring Age	GDM criteria	Method for GDM	Glucose homeostasis	β-Cell morphology and mass	β-Cell function
Male mouse adult offspring (P98–112) C56Bl/6N (Li et al. 2020)	↑ Random BG at GD15 in isolGT (isolated impaired glucose tolerance) mothers 9.6 ± 0.3 mM and in GDM (GDM with resolved hyperglycemia after delivery) mothers 17.4 ± 1.1	Fed HFD (60% kcal) for 4 weeks followed by STZ injection once per day for 3 consecutive days (60 mg/kg BW) and continued on HFD through gestation	↔ Glucose tolerance on NCD Overt glucose intolerance at 16 weeks on HFD		
Male rat adult offspring (15 weeks old) Sprague-Dawley (Agarwal et al. 2019)	↑ Maternal BW ↑ GWG Hyperglycemia Hyperinsulinemia	HFS diet for 6 weeks before mating (45% fat, 35% carbohydrate with 17% sucrose) and continued throughout gestation and suckling	↑ Mild fasting glycemia ↑ Insulinemia ↓ Glucose tolerance on HFS diet	↔ Islet number ↔ Islet mean size ↔ Islet area per pancreas on LF-diet Impaired β-cell expansion on HFS diet	↓ Insulin secretion on LF-diet and HFS diet (in vivo and in vitro)
Rat adult offspring (up to 3 months) Albino rats of Sherman strain (Bihoreau et al. 1986)	Hyperglycemia Hyperinsulinemia	Continuous glucose infusion during last week of pregnancy	↑ Basal glycemia ↔ Basal insulin concentration ↓ Glucose tolerance	↔ Insulin content in 3-month old	↓ Insulin secretion
Rat adult offspring (1–20 months) Wistar (Gauguier et al. 1991)	Mild hyperglycemia (6.5–8 mM)	Continuous glucose infusion during last week of pregnancy	Mild glucose intolerance (1 month) ↑ basal plasma glucose and severe glucose intolerance progressively with age	↔ Insulin concentration (3 months and 8 months)	↓ Insulin secretion (in vivo) ↔ Insulin secretion at 20 months (in vitro)

(continued)

Table 3 (continued)

In the adult offspring

Offspring Age	GDM criteria	Method for GDM	Glucose homeostasis	β-Cell morphology and mass	β-Cell function
Rat adult offspring (6 weeks, 3 months, 6 months) Wistar Goto Kakizaki (Gill-Randall et al. 2004)		Fertilized eggs collected from Wistar female rats transferred into hyperglycemic Goto Kakizaki female rats	↑ Fasting glucose (6 months) ↓ Glucose tolerance (6 months)		
Mouse adult offspring C57BL/6N (Chen et al. 2022)	Non-fasting BG concentrations >300 mg/dL on 30th day post-STZ	Single STZ injection (150 mg/kg BW) followed by IVF and 2-cell embryo transfer into ICR foster mothers	↓ Glucose tolerance (males at 16 weeks; females at 1 year) on NCD ↓ Glucose tolerance on HFD (12 weeks)	↔ Islet size ↔ Islet density ↔ Islet size distribution and morphology ↔ α-/β-cell mass ratio (males at 16 weeks) ↔ Insulin content (males at 16 weeks)	↓ Insulin secretion (males at 16 weeks; females at 24 weeks) (in vitro)

GD, gestational day; STZ, streptozotocin; BG, blood glucose; BW, body weight; GWG, gestational weight gain; LF, low-fat; HFS, high-fat and sucrose; HFD, high-fat diet; NCD, normal chow diet; IVF, in vitro fertilization; ↑, increased or enhanced; ↓, decreased or impaired; ↔, no change or normal relative to control

induced by a combination of HFS diet and STZ, adult male offspring from GDM dams displayed overt glucose intolerance when challenged with an HFD, suggesting an increased vulnerability in response to environmental stress (Li et al. 2020). Furthermore, islets from adult offspring from GDM mothers exhibited impaired GSIS, which was exacerbated by postnatal exposure to HFS diet (Agarwal et al. 2019). Islet gene expression related to inflammation, mitochondrial function, oxidative stress resistance, and protein translation was altered in these offspring from GDM dams.

Embryo transplantation studies further demonstrate the potentially deleterious effects of a diabetic intrauterine environment. Fertilized eggs from Wistar rats, which have a genetically low risk of diabetes, were transferred into either euglycemic Wistar or hyperglycemic Goto Kakizaki (GK) females (Gill-Randall et al. 2004). Compared to the offspring from Wistar gametes transferred into Wistar females, offspring from Wistar gametes in the hyperglycemic GK females were significantly more hyperglycemic at 6 months of age. In vitro fertilization and two-cell embryo transfer of mice oocytes from hyperglycemic mothers to ICR foster mothers produced offspring that developed adult-onset overt glucose intolerance, which was exacerbated when fed a high-fat diet (Chen et al. 2022). This glucose intolerance resulted primarily from impaired GSIS in vivo and in vitro. The expression of TET3 dioxygenase was reduced in oocytes from hyperglycemic females. This downregulation resulted in insufficient demethylation and subsequent hypermethylation at the paternal alleles of several insulin secretion genes, contributing to impaired glucose homeostasis. In summary (summarized in Table 3), these studies demonstrate that early in utero exposure to maternal hyperglycemia may have long-term health consequences that persist into adulthood and increase susceptibility to adverse metabolic health in the adult offspring.

4.5 Potential Mechanisms

Studies in human and animal models have demonstrated that exposure to GDM during a critical development window can alter β-cell structure and function. The studies mentioned above provide supporting evidence that this response by the fetal β-cells potentially sets up long-term β-cell changes that lead to impaired glucose tolerance and a diabetogenic phenotype in the adult offspring. The exact mechanisms involved remain unclear. As previously mentioned, GDM is associated with obesity, prediabetes, and hyperinsulinemia (Alejandro et al. 2020b). Therefore, the mechanisms involved in the programming of the offspring will likely be complex and involve several risk factors. One such mechanism may be maternal inflammation. GDM and obesity are characterized by chronic, low-grade inflammation, which may increase the risk for the offspring to develop disease later in life. In addition to maternal inflammation, the placenta may be a critical mediator between obesity- and GDM-induced inflammation and the intrauterine environment, acting as both a target and a source of inflammatory cytokines (Pantham et al. 2015).

Interestingly, in a rat model of STZ-induced mild diabetes, β-cell mass was reduced in the male adult offspring. This deficit was rescued when the diabetic dams were given olive oil supplementation (Taqui et al. 2020). Taqui et al. propose that the activation of peroxisome proliferator-activated receptors (PPARs), which may have an anti-inflammatory role, may be involved in this rescue (Taqui et al. 2020; Wieser et al. 2008). Growing evidence suggests that exposure to GDM may program the fetus via epigenetic changes (Ruchat et al. 2013; Franzago et al. 2019; Ma et al. 2015). In particular, further investigations into placental epigenetic changes are necessary to better understand how GDM affects the fetus. Alterations in placenta development and function, which can affect nutrient supply and fetal development, are associated with GDM (Franzago et al. 2019). Oxidative stress can also induce epigenetic modifications and may be involved in the programming of the GDM offspring (Tarry-Adkins and Ozanne 2014). Embryos of female Wistar rats administered a single dose of STZ exhibited oxidative DNA damage (Viana et al. 2000). Earlier evidence by Eriksson et al. also provides evidence of reactive oxygen species (ROS) involvement in diabetes (Eriksson and Borg 1991, 1993). Importantly, pancreatic β-cells are more vulnerable to ROS than other tissue cell types and may be more easily damaged by the increase in maternal diabetes-induced ROS (Han et al. 2007; Malaisse et al. 1982; Zhang et al. 1995).

4.6 Paternal Transmission

Because GDM is a maternal condition limited to pregnancy, transmission via the mother and the maternal environment is the focus of much research. In a study tracing the family history of T2D and T1D in women with GDM, the prevalence of diabetes was significantly greater in mothers compared to fathers with significant aggregation of T2D in the maternal-grandmaternal line (Harder et al. 2001).

However, paternal inheritance in T2D is still robust (Fetita et al. 2006; Kasperska-Czyzyk et al. 1996). Therefore, the role of the paternal contribution merits further study, especially because preclinical studies indicate that metabolic phenotypes can be transmitted epigenetically via the father (Schagdarsurengin and Steger 2016; Sales et al. 2017). A study by Wei et al. using a non-genetic prediabetes mouse model found that paternal prediabetes resulted in glucose intolerance and insulin resistance in the offspring with altered gene expression in pancreatic islets (Wei et al. 2014). Another study found that paternal low-dose STZ-induced diabetes was accompanied by insulitis in the offspring's pancreas. Furthermore, a single sub-diabetogenic dose of STZ resulted in impaired insulin secretion in the mouse offspring (Linn et al. 1993). Finally, an additional study found that GDM-exposed male offspring phenotypes were more apparent than those of female offspring in both the F1 and F2 generations. This result suggests that epigenetic modifications in the sperm may be transmitted to the next generation (Ding and Huang 2013).

4.7 Transgenerational Effects

There is also evidence for the transgenerational impact of GDM. Offspring from GDM mothers are at an increased risk of developing GDM during pregnancy (Aerts and Van Assche 2006; Aerts et al. 1990). In addition, in a model of hyperglycemia induced by continuous glucose infusion during the last week of pregnancy, F1 offspring from hyperglycemic rat dams (F0) demonstrated impaired glucose tolerance and insulin secretion in adulthood, consistent with the studies previously discussed (Bihoreau et al. 1986; Gauguier et al. 1990, 1991; Han et al. 2007; Chen et al. 2022).

When mated with control males, the F1 female offspring continued to exhibit impaired glucose tolerance during pregnancy, and their F2 offspring displayed phenotypes consistent with offspring born from diabetic mothers. As adults, these F2 offspring manifested impaired glucose tolerance and insulin secretion, indicating that exposure to maternal GDM during pregnancy increases the diabetogenic susceptibility in the offspring and may also increase the risk for the subsequent generation. However, in the study by Chen et al., the female F1 offspring exposed to maternal hyperglycemia were crossed with control males to assess whether the effects of maternal hyperglycemia persisted into the subsequent generation (F2) (Chen et al. 2022). The F2 generation did not demonstrate impaired glucose tolerance. Thus, the evidence for the transgenerational effects of maternal hyperglycemia exposure is inconsistent and warrants further research.

5 Perspectives and Future Directions

The evidence supporting the fetal origins of obesity and T2D is robust and repeatable across species, including humans. A greater emphasis on mechanism-based and clinical longitudinal studies is needed to uncover and design additional interventions and preventions for T2D. As discussed above, β-cell mass and function, set very early in life, are exquisitely sensitive to nutrient flux during gestation. During the neonatal period, β-cells undergo functional maturation that involves developing a higher threshold for robust GSIS. This process is complex, and the mechanistic cues for β-cell maturation and how it is impacted by maternal nutrition and health during the lactation period are poorly understood. Thus, additional studies are needed to uncover the impact of maternal nutrition on β-cell mass and function during the lactation period. Emerging mechanisms from the studies mentioned above include alteration in epigenetic mechanisms (e.g., DNA methylation and microRNAs) (Park et al. 2008; Alejandro et al. 2014) and the abundance and activity of the nutrient sensors mTORC1 and AMPK (Alejandro et al. 2014; Jaafar et al. 2019; Helman et al. 2020). Indeed, identifying specific miRNAs in islets has advanced our knowledge of the mechanisms of β-cell health programming. However, we still do not know their origins or their β-cell gene targets. The placenta has been suggested as a source of exosomes secreted in the fetal milieu. However, the mechanistic link between placental function and secreted factors and β-cell proliferation and survival regulation has yet to be uncovered. The effects of inflammation in the maternal milieu or in the placenta are also exciting avenues to explore. However, until additional measures are deployed, T2D prevention must start with improving maternal metabolic health with optimum nutrition during pregnancy.

Acknowledgments This work was supported by the National Institutes of Health Grant NIDDK (R21DK112144, R01DK115720, and R56DK121437 to EUA; and T32GM140936 for AW).

Compliance with Ethical Standards Grace Chung declares no conflict of interest. Alicia Wong declares no conflict of interest. Tracy Her declares no conflict of interest. Emilyn U. Alejandro declares no conflict of interest. This chapter does not contain any studies with any human participants or animals performed by any of the authors.

References

Abuzgaia AM, Hardy DB, Arany E (2015) Regulation of postnatal pancreatic Pdx1 and downstream target genes after gestational exposure to protein restriction in rats. Reproduction 149(3):293–303

ACOG (2021) Obesity in pregnancy: ACOG practice bulletin, number 230. Obstet Gynecol 137(6):e128–e144

Aerts L, van Assche FA (1977) Rat foetal endocrine pancreas in experimental diabetes. J Endocrinol 73(2):339–346

Aerts L, Van Assche FA (2006) Animal evidence for the transgenerational development of diabetes mellitus. Int J Biochem Cell Biol 38(5-6):894–903

Aerts L, Holemans K, Van Assche FA (1990) Maternal diabetes during pregnancy: consequences for the offspring. Diabetes Metab Rev 6(3):147–167

Agarwal P et al (2019) Gestational diabetes adversely affects pancreatic islet architecture and function in the male rat offspring. Endocrinology 160(8):1907–1925

Akhaphong B et al (2018) Reduced uterine perfusion pressure causes loss of pancreatic beta cell area but normal function in fetal rat offspring. Am J Physiol Regul Integr Comp Physiol 315(6):R1220–R1231

Akhaphong B et al (2021) Maternal high-fat diet during pre-conception and gestation predisposes adult female offspring to metabolic dysfunction in mice. Front Endocrinol (Lausanne) 12:780300

Alejandro EU et al (2014) Maternal diet-induced microRNAs and mTOR underlie β cell dysfunction in offspring. J Clin Invest 124(10):4395–4410

Alejandro EU et al (2015) Natural history of beta-cell adaptation and failure in type 2 diabetes. Mol Aspects Med 42:19–41

Alejandro EU et al (2017) Overexpression of kinase-dead mTOR impairs glucose homeostasis by regulating insulin secretion and not beta-cell mass. Diabetes 66(8):2150–2162

Alejandro EU et al (2020a) Maternal low-protein diet on the last week of pregnancy contributes to insulin resistance and β-cell dysfunction in the mouse offspring. Am J Physiol Regul Integr Comp Physiol 319(4):R485–R496

Alejandro EU et al (2020b) Gestational diabetes mellitus: a harbinger of the vicious cycle of diabetes. Int J Mol Sci 21(14)

American Diabetes Association (2013) Diagnosis and classification of diabetes mellitus. Diabetes Care 36(Suppl 1):S67–S74

Bartha JL, Romero-Carmona R, Comino-Delgado R (2003) Inflammatory cytokines in intrauterine growth retardation. Acta Obstetricia et Gynecologica Scandinavica 82(12):1099–1102

Bihoreau MT et al (1986) Impaired glucose homeostasis in adult rats from hyperglycemic mothers. Diabetes 35(9):979–984

Blandino-Rosano M et al (2017) Loss of mTORC1 signalling impairs beta-cell homeostasis and insulin processing. Nat Commun 8:16014

Bringhenti I et al (2013) Maternal obesity during the preconception and early life periods alters pancreatic development in early and adult life in male mouse offspring. PLoS One 8(1):e55711

Brown FM, Isganaitis E, James-Todd T (2019) Much to HAPO FUS about: increasing maternal glycemia in pregnancy is associated with worsening childhood glucose metabolism. Diabetes Care 42(3):393–395

Buchanan TA (2001) Pancreatic B-cell defects in gestational diabetes: implications for the pathogenesis and prevention of type 2 diabetes. J Clin Endocrinol Metab 86(3):989–993
Buchanan TA, Xiang AH (2005) Gestational diabetes mellitus. J Clin Invest 115(3):485–491
Buchanan TA et al (1990) Insulin sensitivity and B-cell responsiveness to glucose during late pregnancy in lean and moderately obese women with normal glucose tolerance or mild gestational diabetes. Am J Obstet Gynecol 162(4):1008–1014
Cardell BS (1953) Hypertrophy and hyperplasia of the pancreatic islets in new-born infants. J Pathol Bacteriol 66(2):335–346
Carlson Z et al (2022) Short term changes in dietary fat content and metformin treatment during lactation impact milk composition and mammary gland morphology. J Mammary Gland Biol Neoplasia 27(1):1–18
Cerf ME, Louw J (2010) High fat programming induces glucose intolerance in weanling Wistar rats. Horm Metab Res 42(5):307–310
Cerf ME et al (2005) Islet cell response in the neonatal rat after exposure to a high-fat diet during pregnancy. Am J Physiol Regul Integr Comp Physiol 288(5):R1122–R1128
Cerf ME et al (2006) Hyperglycaemia and reduced glucokinase expression in weanling offspring from dams maintained on a high-fat diet. Br J Nutr 95(2):391–396
Cerf ME et al (2007) Compromised beta-cell development and beta-cell dysfunction in weanling offspring from dams maintained on a high-fat diet during gestation. Pancreas 34(3):347–353
Cerf ME et al (2009) Gestational high-fat programming impairs insulin release and reduces Pdx-1 and glucokinase immunoreactivity in neonatal Wistar rats. Metabolism 58(12):1787–1792
Chamson-Reig A et al (2006) Altered pancreatic morphology in the offspring of pregnant rats given reduced dietary protein is time and gender specific. J Endocrinol 191(1):83–92
Chen B et al (2022) Maternal inheritance of glucose intolerance via oocyte TET3 insufficiency. Nature 605(7911):761–766
Chivese T, Norris SA, Levitt NS (2019) Progression to type 2 diabetes mellitus and associated risk factors after hyperglycemia first detected in pregnancy: A cross-sectional study in Cape Town, South Africa. PLoS Med 16(9):e1002865
Choi J et al (2011) Emergence of insulin resistance in juvenile baboon offspring of mothers exposed to moderate maternal nutrient reduction. Am J Physiol Regul Integr Comp Physiol 301(3):R757–R762
Clausen TD et al (2008) High prevalence of type 2 diabetes and pre-diabetes in adult offspring of women with gestational diabetes mellitus or type 1 diabetes: the role of intrauterine hyperglycemia. Diabetes Care 31(2):340–346
Clausen TD et al (2009) Overweight and the metabolic syndrome in adult offspring of women with diet-treated gestational diabetes mellitus or type 1 diabetes. J Clin Endocrinol Metab 94(7):2464–2470
Claycombe-Larson KG, Bundy AN, Roemmich JN (2020) Paternal high-fat diet and exercise regulate sperm miRNA and histone methylation to modify placental inflammation, nutrient transporter mRNA expression and fetal weight in a sex-dependent manner. J Nutr Biochem 81:108373
Comstock SM et al (2012) High-fat diet consumption during pregnancy and the early post-natal period leads to decreased α cell plasticity in the nonhuman primate. Mol Metab 2(1):10–22
Coustan DR (2016) Recurrent GDM and the development of type 2 diabetes have similar risk factors. Endocrine 53(3):624–625
Cox AR et al (2010) The effects of low protein during gestation on mouse pancreatic development and beta cell regeneration. Pediatr Res 68(1):16–22
Dabelea D et al (2000) Intrauterine exposure to diabetes conveys risks for type 2 diabetes and obesity: a study of discordant sibships. Diabetes 49(12):2208–2211
D'Agostino AN, Bahn RC (1963) A histopathologic study of the pancreas of infants of diabetic mothers. Diabetes 12:327–331
Dahri S et al (1991) Islet function in offspring of mothers on low-protein diet during gestation. Diabetes 40(Suppl 2):115–120

Damm P et al (2016) Gestational diabetes mellitus and long-term consequences for mother and offspring: a view from Denmark. Diabetologia 59(7):1396–1399
Dearden L, Bouret SG, Ozanne SE (2018) Sex and gender differences in developmental programming of metabolism. Mol Metab 15:8–19
Dearden L et al (2020) Maternal obesity causes fetal hypothalamic insulin resistance and disrupts development of hypothalamic feeding pathways. Mol Metab 42:101079
de Rooij SR et al (2006a) Impaired insulin secretion after prenatal exposure to the Dutch famine. Diabetes Care 29(8):1897–1901
de Rooij SR et al (2006b) Glucose tolerance at age 58 and the decline of glucose tolerance in comparison with age 50 in people prenatally exposed to the Dutch famine. Diabetologia 49(4):637–643
de Souza AH et al (2020) Intra-islet GLP-1, but not CCK, is necessary for β-cell function in mouse and human islets. Sci Rep 10(1):2823
Devlieger R, Casteels K, Van Assche FA (2008) Reduced adaptation of the pancreatic B cells during pregnancy is the major causal factor for gestational diabetes: current knowledge and metabolic effects on the offspring. Acta Obstet Gynecol Scand 87(12):1266–1270
Diaz P et al (2015) Increased placental fatty acid transporter 6 and binding protein 3 expression and fetal liver lipid accumulation in a mouse model of obesity in pregnancy. Am J Physiol Regul Integr Comp Physiol 309(12):R1569–R1577
Ding GL, Huang HF (2013) Paternal transgenerational glucose intolerance with epigenetic alterations in second generation offspring of GDM. Asian J Androl 15(4):451–452
Dor Y et al (2004) Adult pancreatic beta-cells are formed by self-duplication rather than stem-cell differentiation. Nature 429(6987):41–46
Duale N et al (2014) Impaired sperm chromatin integrity in obese mice. Andrology 2(2):234–243
Dumortier O et al (2007) Different mechanisms operating during different critical time-windows reduce rat fetal beta cell mass due to a maternal low-protein or low-energy diet. Diabetologia 50(12):2495–2503
Dumortier O et al (2014) Maternal protein restriction leads to pancreatic failure in offspring: role of misexpressed microRNA-375. Diabetes 63(10):3416–3427
Duttaroy AK, Basak S (2020) Maternal dietary fatty acids and their roles in human placental development. Prostaglandins Leukot Essent Fatty Acids 155:102080
Eades CE et al (2015) Progression from gestational diabetes to type 2 diabetes in one region of Scotland: an observational follow-up study. BMC Pregnancy Childbirth 15:11
Elghazi L et al (2017) Role of nutrients and mTOR signaling in the regulation of pancreatic progenitors development. Mol Metab 6(6):560–573
Elsakr JM et al (2019) Maternal Western-style diet affects offspring islet composition and function in a non-human primate model of maternal over-nutrition. Mol Metab 25:73–82
Eriksson UJ, Borg LA (1991) Protection by free oxygen radical scavenging enzymes against glucose-induced embryonic malformations in vitro. Diabetologia 34(5):325–331
Eriksson UJ, Borg LA (1993) Diabetes and embryonic malformations. Role of substrate-induced free-oxygen radical production for dysmorphogenesis in cultured rat embryos. Diabetes 42(3):411–419
Eriksson JG et al (2014) Long-term consequences of maternal overweight in pregnancy on offspring later health: findings from the Helsinki Birth Cohort Study. Ann Med 46(6):434–438
Eriksson JG et al (2015) Maternal weight in pregnancy and offspring body composition in late adulthood: findings from the Helsinki Birth Cohort Study (HBCS). Ann Med 47(2):94–99
Fall CH et al (1998) Size at birth, maternal weight, and type 2 diabetes in South India. Diabet Med 15(3):220–227
Fernandez-Twinn DS, Ozanne SE (2010) Early life nutrition and metabolic programming. Ann N Y Acad Sci 1212:78–96
Ferrannini E et al (2005) beta-Cell function in subjects spanning the range from normal glucose tolerance to overt diabetes: a new analysis. J Clin Endocrinol Metab 90(1):493–500

Fetita LS et al (2006) Consequences of fetal exposure to maternal diabetes in offspring. J Clin Endocrinol Metab 91(10):3718–3724
Filhoulaud G, Guillemain G, Scharfmann R (2009) The hexosamine biosynthesis pathway is essential for pancreatic beta cell development. J Biol Chem 284(36):24583–24594
Ford SP et al (2009) Maternal obesity accelerates fetal pancreatic beta-cell but not alpha-cell development in sheep: prenatal consequences. Am J Physiol Regul Integr Comp Physiol 297(3):R835–R843
Franzago M et al (2019) Nutrigenetics, epigenetics and gestational diabetes: consequences in mother and child. Epigenetics 14(3):215–235
Fridolf T et al (1991) GLP-1 and GLP-1(7-36) amide: influences on basal and stimulated insulin and glucagon secretion in the mouse. Pancreas 6(2):208–215
Gao T et al (2014) Pdx1 maintains beta cell identity and function by repressing an alpha cell program. Cell Metab 19(2):259–271
Gauguier D et al (1990) Inheritance of diabetes mellitus as consequence of gestational hyperglycemia in rats. Diabetes 39(6):734–739
Gauguier D et al (1991) Insulin secretion in adult rats after intrauterine exposure to mild hyperglycemia during late gestation. Diabetes 40(Suppl 2):109–114
Gawlińska K et al (2021) Relationship of maternal high-fat diet during pregnancy and lactation to offspring health. Nutr Rev 79(6):709–725
Geraghty AA et al (2015) Nutrition during pregnancy impacts offspring's epigenetic status-evidence from human and animal studies. Nutr Metab Insights 8(Suppl 1):41–47
Gill-Randall R et al (2004) Type 2 diabetes mellitus—genes or intrauterine environment? An embryo transfer paradigm in rats. Diabetologia 47(8):1354–1359
Govoni KE, Reed SA, Zinn SA (2019) Cell biology symposium: metabolic responses to stress: from animal to cell: Poor maternal nutrition during gestation: effects on offspring whole-body and tissue-specific metabolism in livestock species1,2. J Anim Sci 97(7):3142–3152
Graus-Nunes F et al (2015) Pregestational maternal obesity impairs endocrine pancreas in male F1 and F2 progeny. Nutrition 31(2):380–387
Gregg BE et al (2012) Formation of a human beta-cell population within pancreatic islets is set early in life. J Clin Endocrinol Metab 97(9):3197–3206
Guillemain G et al (2007) Glucose is necessary for embryonic pancreatic endocrine cell differentiation. J Biol Chem 282(20):15228–15237
Haase TN et al (2013) Growth arrest specific protein (GAS) 6: a role in the regulation of proliferation and functional capacity of the perinatal rat beta cell. Diabetologia 56(4):763–773
Han J et al (2007) Rat maternal diabetes impairs pancreatic beta-cell function in the offspring. Am J Physiol Endocrinol Metab 293(1):E228–E236
Harder T et al (2001) Maternal and paternal family history of diabetes in women with gestational diabetes or insulin-dependent diabetes mellitus type I. Gynecol Obstet Invest 51(3):160–164
Helman A et al (2020) A nutrient-sensing transition at birth triggers glucose-responsive insulin secretion. Cell Metab 31(5):1004–1016 e5
Heltemes A et al (2010) Chronic placental ischemia alters amniotic fluid milieu and results in impaired glucose tolerance, insulin resistance and hyperleptinemia in young rats. Exp Biol Med (Maywood) 235(7):892–899
Herath H, Herath R, Wickremasinghe R (2017) Gestational diabetes mellitus and risk of type 2 diabetes 10 years after the index pregnancy in Sri Lankan women—A community based retrospective cohort study. PLoS One 12(6):e0179647
Hoet JJ (1991) Islet function in offspring of mothers on low-protein diet during gestation. Diabetes 40:115–120
Holemans K, Aerts L, Van Assche FA (2003) Lifetime consequences of abnormal fetal pancreatic development. J Physiol 547(Pt 1):11–20
Homko C et al (2001) Insulin secretion during and after pregnancy in patients with gestational diabetes mellitus. J Clin Endocrinol Metab 86(2):568–573
Huang YH et al (2017) Maternal high-fat diet impairs glucose metabolism, β-cell function and proliferation in the second generation of offspring rats. Nutr Metab (Lond) 14:67

Hull RL et al (2005) Dietary-fat-induced obesity in mice results in beta cell hyperplasia but not increased insulin release: evidence for specificity of impaired beta cell adaptation. Diabetologia 48(7):1350–1358

Jaafar R et al (2019) mTORC1 to AMPK switching underlies beta-cell metabolic plasticity during maturation and diabetes. J Clin Invest 129(10):4124–4137

Jones HN et al (2009) High-fat diet before and during pregnancy causes marked up-regulation of placental nutrient transport and fetal overgrowth in C57/BL6 mice. FASEB J 23(1):271–278

Kasperska-Czyzyk T et al (1996) Difference in the influence of maternal and paternal NIDDM on pancreatic beta-cell activity and blood lipids in normoglycaemic non-diabetic adult offspring. Diabetologia 39(7):831–837

Keleher MR et al (2018) Maternal high-fat diet associated with altered gene expression, DNA methylation, and obesity risk in mouse offspring. PLoS One 13(2):e0192606

Kelly AC, Powell TL, Jansson T (2020) Placental function in maternal obesity. Clin Sci (Lond) 134(8):961–984

Kelstrup L et al (2013) Insulin resistance and impaired pancreatic β-cell function in adult offspring of women with diabetes in pregnancy. J Clin Endocrinol Metab 98(9):3793–3801

Kervran A, Guillaume M, Jost A (1978) The endocrine pancreas of the fetus from diabetic pregnant rat. Diabetologia 15(5):387–393

King R et al (2019) Offspring of mice exposed to a low-protein diet in utero demonstrate changes in mTOR signaling in pancreatic islets of langerhans, associated with altered glucagon and insulin expression and a lower beta-cell mass. Nutrients 11(3)

Kirigiti MA et al (2020) Effects of pre- and postnatal protein restriction on maternal and offspring metabolism in the nonhuman primate. Am J Physiol Regul Integr and Comp Physiol 318(5):R929–R939

Kongsted AH et al (2014) Programming of glucose-insulin homoeostasis: long-term consequences of pre-natal versus early post-natal nutrition insults. Evidence from a sheep model. Acta Physiol (Oxford, England) 210(1):84–98

Ktorza A et al (1981) Hyperglycaemia induced by glucose infusion in the unrestrained pregnant rat during the last three days of gestation: metabolic and hormonal changes in the mother and the fetuses. Diabetologia 21(6):569–574

Kühl C (1991) Insulin secretion and insulin resistance in pregnancy and GDM. Implications for diagnosis and management. Diabetes 40(Suppl 2):18–24

Kunz LH, King JC (2007) Impact of maternal nutrition and metabolism on health of the offspring. Semin Fetal Neonatal Med 12(1):71–77

Laitinen J, Power C, Järvelin MR (2001) Family social class, maternal body mass index, childhood body mass index, and age at menarche as predictors of adult obesity. Am J Clin Nutr 74(3):287–294

Leturque A et al (1987) Effect of feeding a high-fat diet during pregnancy on glucose metabolism in the rat. Metabolism 36(1):66–70

Li HY et al (2020) A mouse model of gestation-specific transient hyperglycemia for translational studies. J Endocrinol 244(3):501–510

Limesand SW et al (2005) Diminished β-cell replication contributes to reduced β-cell mass in fetal sheep with intrauterine growth restriction. Am J Physiol Regul Integr Comp Physiol 288(5):R1297–R1305

Limesand SW et al (2007) Increased insulin sensitivity and maintenance of glucose utilization rates in fetal sheep with placental insufficiency and intrauterine growth restriction. Am J Physiol Endocrinol Metab 293(6):E1716–E1725

Linn T et al (1993) Spontaneous glucose intolerance in the progeny of low dose streptozotocin-induced diabetic mice. Diabetologia 36(12):1245–1251

Liu Z et al (2020) Pregestational diet transition to normal-fat diet avoids the deterioration of pancreatic β-cell function in male offspring induced by maternal high-fat diet. J Nutr Biochem 86:108495

Lowe WL et al (2019) Hyperglycemia and adverse pregnancy outcome follow-up study (HAPO FUS): maternal gestational diabetes mellitus and childhood glucose metabolism. Diabetes Care 42(3):372–380

Lowensohn RI, Stadler DD, Naze C (2016) Current concepts of maternal nutrition. Obstet Gynecol Surv 71(7):413–426

Lucas A (1998) Programming by early nutrition: an experimental approach. J Nutr 128(2 Suppl):401S–406S

Lucas A, Fewtrell MS, Cole TJ (1999) Fetal origins of adult disease-the hypothesis revisited. BMJ 319(7204):245–249

Ma RC et al (2015) Maternal diabetes, gestational diabetes and the role of epigenetics in their long term effects on offspring. Prog Biophys Mol Biol 118(1-2):55–68

Macfarlane CM, Tsakalakos N (1988) The extended Pedersen hypothesis. Clin Physiol Biochem 6(2):68–73

Malaisse WJ et al (1982) Determinants of the selective toxicity of alloxan to the pancreatic B cell. Proc Natl Acad Sci U S A 79(3):927–930

Malcolm JC et al (2006) Glucose tolerance of offspring of mother with gestational diabetes mellitus in a low-risk population. Diabet Med 23(5):565–570

Masuyama H, Hiramatsu Y (2014) Additive effects of maternal high fat diet during lactation on mouse offspring. PLoS One 9(3):e92805

Moede T, Leibiger IB, Berggren PO (2020) Alpha cell regulation of beta cell function. Diabetologia 63(10):2064–2075

Mohan R, Baumann D, Alejandro EU (2018) Fetal undernutrition, placental insufficiency, and pancreatic β-cell development programming in utero. Am J Physiol Regul Integr Comp Physiol 315(5):R867–R878

Morgan HL et al (2021) Paternal low protein diet and the supplementation of methyl-donors impact fetal growth and placental development in mice. Placenta 103:124–133

Newsome CA et al (2003) Is birth weight related to later glucose and insulin metabolism?—a systematic review. Diabet Med 20(5):339–348

Ng SF et al (2010) Chronic high-fat diet in fathers programs β-cell dysfunction in female rat offspring. Nature 467(7318):963–966

Nicholas LM et al (2020) Exposure to maternal obesity programs sex differences in pancreatic islets of the offspring in mice. Diabetologia 63(2):324–337

Noctor E, Dunne FP (2015) Type 2 diabetes after gestational diabetes: The influence of changing diagnostic criteria. World J Diab 6(2):234–244

Ogata ES et al (1980) Perinatal islet function in gestational diabetes: assessment by cord plasma C-peptide and amniotic fluid insulin. Diabetes Care 3(3):425–429

Orskov C, Holst JJ, Nielsen OV (1988) Effect of truncated glucagon-like peptide-1 [proglucagon-(78-107) amide] on endocrine secretion from pig pancreas, antrum, and nonantral stomach. Endocrinology 123(4):2009–2013

Ou XH, Zhu CC, Sun SC (2019) Effects of obesity and diabetes on the epigenetic modification of mammalian gametes. J Cell Physiol 234(6):7847–7855

Ozanne SE, Sandovici I, Constância M (2011) Maternal diet, aging and diabetes meet at a chromatin loop. Aging 3(5):548–554

Palou M et al (2010) Sexual dimorphism in the lasting effects of moderate caloric restriction during gestation on energy homeostasis in rats is related with fetal programming of insulin and leptin resistance. Nutr Metab 7(1):69

Pang H et al (2021) Gestational high-fat diet impaired demethylation of Pparα and induced obesity of offspring. J Cell Mol Med 25(12):5404–5416

Pantham P, Aye IL, Powell TL (2015) Inflammation in maternal obesity and gestational diabetes mellitus. Placenta 36(7):709–715

Park JH et al (2008) Development of type 2 diabetes following intrauterine growth retardation in rats is associated with progressive epigenetic silencing of Pdx1. J Clin Invest 118(6):2316–2324

Parrettini S, Caroli A, Torlone E (2020) Nutrition and metabolic adaptations in physiological and complicated pregnancy: focus on obesity and gestational diabetes. Front Endocrinol (Lausanne) 11:611929
Pasek RC, Gannon M (2013) Advancements and challenges in generating accurate animal models of gestational diabetes mellitus. Am J Physiol Endocrinol Metab 305(11):E1327–E1338
Patel S et al (2016) Association of maternal diabetes/glycosuria and pre-pregnancy body mass index with offspring indicators of non-alcoholic fatty liver disease. BMC Pediatr 16:47
Peng H et al (2021) Maternal high-fat diet disrupted one-carbon metabolism in offspring, contributing to nonalcoholic fatty liver disease. Liver Int 41(6):1305–1319
Petrik J et al (1999a) A low protein diet alters the balance of islet cell replication and apoptosis in the fetal and neonatal rat and is associated with a reduced pancreatic expression of insulin-like growth factor-II. Endocrinology 140(10):4861–4873
Petrik J et al (1999b) Overexpression of insulin-like growth factor-II in transgenic mice is associated with pancreatic islet cell hyperplasia. Endocrinology 140(5):2353–2363
Plagemann A et al (1997) Glucose tolerance and insulin secretion in children of mothers with pregestational IDDM or gestational diabetes. Diabetologia 40(9):1094–1100
Plows JF et al (2018) The pathophysiology of gestational diabetes mellitus. Int J Mol Sci 19(11)
Qi L et al (2022) Dietary curcumin supplementation ameliorates placental inflammation in rats with intra-uterine growth retardation by inhibiting the NF-κB signaling pathway. J Nutr Biochem 104:108973
Qiao L et al (2015) Maternal high-fat feeding increases placental lipoprotein lipase activity by reducing SIRT1 expression in mice. Diabetes 64(9):3111–3120
Qiao L et al (2019) Prolonged prepregnant maternal high-fat feeding reduces fetal and neonatal blood glucose concentrations by enhancing fetal β-cell development in C57BL/6 mice. Diabetes 68(8):1604–1613
Rachdi L et al (2012) L-leucine alters pancreatic β-cell differentiation and function via the mTor signaling pathway. Diabetes 61(2):409–417
Rando OJ, Simmons RA (2015) I'm eating for two: parental dietary effects on offspring metabolism. Cell 161(1):93–105
Ravelli GP, Stein ZA, Susser MW (1976) Obesity in young men after famine exposure in utero and early infancy. N Engl J Med 295(7):349–353
Rayanagoudar G et al (2016) Quantification of the type 2 diabetes risk in women with gestational diabetes: a systematic review and meta-analysis of 95,750 women. Diabetologia 59(7):1403–1411
Reiher H et al (1983) Age-dependent insulin secretion of the endocrine pancreas in vitro from fetuses of diabetic and nondiabetic patients. Diabetes Care 6(5):446–451
Reusens-Billen B et al (1984) Cell proliferation in pancreatic islets of rat fetuses and neonates from normal and diabetic mothers. An in vitro and in vivo study. Horm Metab Res 16(11):565–571
Reynolds CM et al (2015) Maternal high fat and/or salt consumption induces sex-specific inflammatory and nutrient transport in the rat placenta. Physiol Rep 3(5)
Rodriguez UA et al (2021) Conversion of α-cells to β-cells in the postpartum mouse pancreas involves Lgr5 progeny. Diabetes 70(7):1508–1518
Rodríguez-Trejo A et al (2012) Developmental programming of neonatal pancreatic β-cells by a maternal low-protein diet in rats involves a switch from proliferation to differentiation. Am J Physiol Endocrinol Metab 302(11):E1431–E1439
Rodriguez-Trejo A et al (2012) Developmental programming of neonatal pancreatic beta-cells by a maternal low-protein diet in rats involves a switch from proliferation to differentiation. Am J Physiol Endocrinol Metab 302(11):E1431–E1439
Root KM, Akhaphong B, Cedars MA, Molin AM, Huchthausen ME, Laule CF, Regal RR, Alejandro EU, Regal JF (2022) Critical role for macrophages in the developmental programming of pancreatic β-cell area in offspring of hypertensive pregnancies. Diabetes 71(12):2597–2611
Roseboom TJ et al (2003) Perceived health of adults after prenatal exposure to the Dutch famine. Paediatr Perinat Epidemiol 17(4):391–397

Roseboom T, de Rooij S, Painter R (2006) The Dutch famine and its long-term consequences for adult health. Early Hum Dev 82(8):485–491
Ruchat SM et al (2013) Gestational diabetes mellitus epigenetically affects genes predominantly involved in metabolic diseases. Epigenetics 8(9):935–943
Saengnipanthkul S et al (2021) Maternal exposure to high-fat diet during pregnancy and lactation predisposes normal weight offspring mice to develop hepatic inflammation and insulin resistance. Physiol Rep 9(6):e14811
Sales VM, Ferguson-Smith AC, Patti ME (2017) Epigenetic mechanisms of transmission of metabolic disease across generations. Cell Metab 25(3):559–571
Sandovici I et al (2012) Placental adaptations to the maternal–fetal environment: implications for fetal growth and developmental programming. Reprod BioMed Online 25(1):68–89
Sandovici I et al (2013) Developmental and environmental epigenetic programming of the endocrine pancreas: consequences for type 2 diabetes. Cell Mol Life Sci 70(9):1575–1595
Schack-Nielsen L et al (2010) Gestational weight gain in relation to offspring body mass index and obesity from infancy through adulthood. Int J Obes (Lond) 34(1):67–74
Schagdarsurengin U, Steger K (2016) Epigenetics in male reproduction: effect of paternal diet on sperm quality and offspring health. Nat Rev Urol 13(10):584–595
Schellong K et al (2020) Maternal but not paternal high-fat diet (HFD) exposure at conception predisposes for 'diabesity' in offspring generations. Int J Environ Res Public Health 17(12)
Schmidt WE, Siegel EG, Creutzfeldt W (1985) Glucagon-like peptide-1 but not glucagon-like peptide-2 stimulates insulin release from isolated rat pancreatic islets. Diabetologia 28(9):704–707
Schneider S et al (2012) The prevalence of gestational diabetes in advanced economies. J Perinat Med 40(5):511–520
Scholtens DM et al (2019) Hyperglycemia and adverse pregnancy outcome follow-up study (HAPO FUS): maternal glycemia and childhood glucose metabolism. Diabetes Care 42(3):381–392
Silverman BL et al (1995) Impaired glucose tolerance in adolescent offspring of diabetic mothers. Relationship to fetal hyperinsulinism. Diabetes Care 18(5):611–617
Simmons RA, Templeton LJ, Gertz SJ (2001) Intrauterine growth retardation leads to the development of type 2 diabetes in the rat. Diabetes 50(10):2279–2286
Snoeck A, Remacle C, Reusens B, Hoet JJ (1990) Effect of a low protein diet during pregnancy on the fetal rat endocrine pancreas. Biol Neonate 57:107–118
Song L et al (2017) Prenatal high-fat diet alters placental morphology, nutrient transporter expression, and mtorc1 signaling in rat. Obesity (Silver Spring) 25(5):909–919
Sparre T et al (2003) Intrauterine programming of fetal islet gene expression in rats—effects of maternal protein restriction during gestation revealed by proteome analysis. Diabetologia 46(11):1497–1511
Stanger BZ, Tanaka AJ, Melton DA (2007) Organ size is limited by the number of embryonic progenitor cells in the pancreas but not the liver. Nature 445(7130):886–891
Steinke J, Driscoll SG (1965) The extractable insulin content of pancreas from fetuses and infants of diabetic and control mothers. Diabetes 14(9):573–578
Stoffers DA, Thomas MK, Habener JF (1997) Homeodomain protein IDX-1: a master regulator of pancreas development and insulin gene expression. Trends Endocrinol Metab 8(4):145–151
Su Y et al (2016) Maternal low protein isocaloric diet suppresses pancreatic beta-cell proliferation in mouse offspring via miR-15b. Endocrinology 157(12):4782–4793
Taqui B et al (2020) Addition of olive oil to diet of rats with mild pre-gestational diabetes impacts offspring beta-cell development. J Endocrinol 246(2):175–187
Tarry-Adkins JL, Ozanne SE (2014) The impact of early nutrition on the ageing trajectory. Proc Nutr Soc 73(2):289–301
Terashima M et al (2015) Effect of high fat diet on paternal sperm histone distribution and male offspring liver gene expression. Epigenetics 10(9):861–871
Thorel F et al (2010) Conversion of adult pancreatic alpha-cells to beta-cells after extreme beta-cell loss. Nature 464(7292):1149–1154

Utzschneider KM et al (2009) Oral disposition index predicts the development of future diabetes above and beyond fasting and 2-h glucose levels. Diabetes Care 32(2):335–341
Viana M et al (2000) Oxidative damage in pregnant diabetic rats and their embryos. Free Radic Biol Med 29(11):1115–1121
Vipin VA, Blesson CS, Yallampalli C (2022) Maternal low protein diet and fetal programming of lean type 2 diabetes. World J Diab 13(3):185–202
Wang B et al (2022) Paternal high-fat diet altered sperm 5′tsRNA-Gly-GCC is associated with enhanced gluconeogenesis in the offspring. Front Mol Biosci 9:857875
Watkins AJ, Sinclair KD (2014) Paternal low protein diet affects adult offspring cardiovascular and metabolic function in mice. Am J Physiol Heart Circ Physiol 306(10):H1444–H1452
Wei Y et al (2014) Paternally induced transgenerational inheritance of susceptibility to diabetes in mammals. Proc Natl Acad Sci U S A 111(5):1873–1878
Wieser F et al (2008) PPAR action in human placental development and pregnancy and its complications. PPAR Res 2008:527048
Yang J et al (2004) Leucine culture reveals that ATP synthase functions as a fuel sensor in pancreatic β-cells. J Biol Chem 279(52):53915–53923
Yang J et al (2010) Leucine metabolism in regulation of insulin secretion from pancreatic beta cells. Nutr Rev 68(5):270–279
Yuan Q et al (2011) Postnatal pancreatic islet beta cell function and insulin sensitivity at different stages of lifetime in rats born with intrauterine growth retardation. PLoS One 6(10):e25167
Zhang H, Ollinger K, Brunk U (1995) Insulinoma cells in culture show pronounced sensitivity to alloxan-induced oxidative stress. Diabetologia 38(6):635–641
Zhang Y et al (2019a) Sperm RNA code programmes the metabolic health of offspring. Nat Rev Endocrinol 15(8):489–498
Zhang Q et al (2019b) A maternal high-fat diet induces DNA methylation changes that contribute to glucose intolerance in offspring. Front Endocrinol (Lausanne) 10:871
Zheng J et al (2020) Maternal high-fat diet regulates glucose metabolism and pancreatic β cell phenotype in mouse offspring at weaning. PeerJ 8:e9407
Zhu Y, Zhang C (2016) Prevalence of gestational diabetes and risk of progression to type 2 diabetes: a global perspective. Curr Diab Rep 16(1):7

GPSR Compliance

The European Union's (EU) General Product Safety Regulation (GPSR) is a set of rules that requires consumer products to be safe and our obligations to ensure this.

If you have any concerns about our products, you can contact us on ProductSafety@springernature.com

In case Publisher is established outside the EU, the EU authorized representative is:

Springer Nature Customer Service Center GmbH
Europaplatz 3
69115 Heidelberg, Germany

Batch number: 10732628

Printed by Printforce, the Netherlands